KEYNOTE guide to
topics in your course

CLIFFS KEYNOTE REVIEWS

Physics

by

ANDRES MANRING

CLIFF'S NOTES, INC. • LINCOLN, NEBRASKA 68501

ISBN   0-8220-1735-0

© Copyright 1969

BY CLIFF'S NOTES, INC.

L. C. Catalogue Card Number: 73-89834

Printed in the United States of America

CONTENTS

TO THE STUDENT

This KEYNOTE is a flexible study aid designed to help you REVIEW YOUR COURSE QUICKLY and USE YOUR TIME TO STUDY ONLY MATERIAL YOU DON'T KNOW.

FOR GENERAL REVIEW

Take the SELF-TEST on the first page of any topic and turn the page to check your answers.

Read the EXPLANATIONS of any questions you answered incorrectly.

If you are satisfied with your understanding of the material, move on to another topic.

If you feel that you need further review, read the column of BASIC FACTS. For a more detailed discussion of the material, read the column of ADDITIONAL INFORMATION.

FOR QUICK REVIEW

Read the column of BASIC FACTS for a rapid review of the essentials of a topic and then take the SELF-TEST if you want to test your understanding of the material.

ADDITIONAL HELP FOR EXAMS

Review terms in the DICTIONARY-INDEX.
Test yourself by taking the sample FINAL EXAM.

MATHEMATICS REVIEW

SELF-TEST

DIRECTIONS. For all questions, write your answers on the numbered line in the right margin (of page 3). Then turn page 1 so that you can see the printed answers next to your answers. Study the explanations on page 3 for any questions you missed. Review material is given under Basic Facts and Additional Information.

1. Given three vectors: $\mathbf{a} = 4\mathbf{i} + 3\mathbf{j} - 5\mathbf{k}$, $\mathbf{b} = 2\mathbf{i} + 4\mathbf{j} + 6\mathbf{k}$, and $\mathbf{c} = -\mathbf{j} + 5\mathbf{k}$.
   a What are the magnitudes of $\mathbf{a}$, $\mathbf{b}$, and $\mathbf{c}$?
   b Find $\mathbf{a} + \mathbf{b} + \mathbf{c}$.
   c Find $\mathbf{a} - \mathbf{b}$.
   d Find $\mathbf{a} \cdot \mathbf{b}$.
   e Find $\mathbf{a} \times \mathbf{b}$ and $\mathbf{b} \times \mathbf{a}$.

2. Given a vector in two dimensions: $\mathbf{a} = \mathbf{i} + 2\mathbf{j}$, what are the components of $\mathbf{a}$ in terms of the $\sin \theta$ and $\cos \theta$?

3. Express the following values using scientific notation:
   a 9,770,000,000.
   b 765.32.
   c 0.0013.
   d 1.210.
   e 0.0000035.

1 a
  b
  c
  d
  e

2

3 a
  b
  c
  d
  e

1

BASIC FACTS

Systems of Units. The commonly used systems of units are the **mks** (meter-kilogram-second), **cgs** (centimeter-gram-second), and **fps** (foot-pound-second) systems. In the fps system, the pound is a unit of weight, not of mass. In this system the unit for mass is the slug: 1 slug = 1 pound/g, where g = 32 ft/sec^2 is the acceleration of gravity. The fundamental units of a system are *length, time, temperature, charge, and mass.*

Scalars and Vectors. A scalar is a quantity that has only magnitude; therefore it can be represented by a *number* and a *unit*. Scalars are used to measure density, length, time, temperature, and energy; since they are numbers, the ordinary rules of algebra apply to them.

A vector represents *displacement* relative to a coordinate system; therefore, a vector has magnitude, direction, and *sense*. Vectors are usually pictured as arrows; thus, the magnitude of a vector is the length of the arrow, its direction is its angular position with respect to the axes, and the sense is the position of the arrowhead, or *tip*, with respect to the *tail*. Hence, if the tip and tail are interchanged, the sense is reversed. The arrow is always straight and connects only two points, the initial (tail) and terminal (tip) points, even though these points may be in a curved path.

In mathematical notation, vectors are variously symbolized as **boldface** (heavy) letters...**a, b, c,**..., or as letters with arrows above them...$\vec{a}, \vec{b}, \vec{c}$,.... Here, boldface letters will be used.

Frequently, vectors which represent forces must be *resolved* into component vectors, or simply **components**. Fig. 1-1 shows parallelogram $OABC$ with vectors $\mathbf{a} = \overrightarrow{OB}$, $\mathbf{b} = \overrightarrow{OA}$, and $\mathbf{c} = \overrightarrow{OC}$, where **b** and **c** are the components of **a**;

(Continued on page 4)

ADDITIONAL INFORMATION

The magnitude of a vector **a** is its *absolute value* $|\mathbf{a}| = a$, which is a scalar. The *unit components* of a vector lie along axes x, y, and z, and are designated as **i, j,** and **k**, respectively. Thus a vector $\mathbf{a} = a_x\mathbf{i} + a_y\mathbf{j} + a_z\mathbf{k}$, where a_x, a_y, and a_z are the lengths (scalar multiples) of **i, j,** and **k** (Fig. 1-3). Also, in two-dimen-

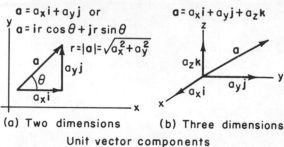

(a) Two dimensions   (b) Three dimensions
Unit vector components

Fig. 1-3

sional space, $a_x = a\cos\theta$, $a_y = a\sin\theta$, and $a_y/a_x = \tan\theta$, where $a = \sqrt{a_x^2 + a_y^2}$. Vectors may also be added by adding their components, since each vector is the sum of its components.

The two methods for multiplying vectors are the **scalar** or **dot product** and the **vector** or **cross product**. The dot product of two vectors, $\mathbf{a} \cdot \mathbf{b} = ab\cos\theta$ represents a *scalar* value.   The dot products of the unit components are, $\mathbf{i} \cdot \mathbf{i} = \mathbf{j} \cdot \mathbf{j} = \mathbf{k} \cdot \mathbf{k} = 1$, but $\mathbf{i} \cdot \mathbf{j} = \mathbf{j} \cdot \mathbf{i} = \mathbf{i} \cdot \mathbf{k} = \mathbf{k} \cdot \mathbf{i} = \mathbf{j} \cdot \mathbf{k} = \mathbf{k} \cdot \mathbf{j} = 0$, since **i, j,** and **k** are mutually perpendicular. Hence, $|\mathbf{a}| = \sqrt{(a_x\mathbf{i})^2 + (a_y\mathbf{j})^2 + (a_z\mathbf{k})^2} = \sqrt{a_x^2 + a_y^2 + a_z^2}$.

The cross product of two vectors, $\mathbf{a} \times \mathbf{b}$ (**a** cross **b**), lying in the same plane is a vector **c** (Fig. 1-4) perpendicular to both **a** and **b** in a direction given by the right-hand rule: make a fist with the right hand and

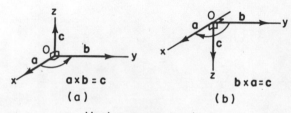

$\mathbf{a} \times \mathbf{b} = \mathbf{c}$ $\mathbf{b} \times \mathbf{a} = \mathbf{c}$
(a)   (b)

Vector cross product

Fig. 1-4

let thumb point upward. The fingers are then made to point from **a** to **b**, and the thumb will indicate the

(Continued on page 4)

EXPLANATIONS

1. a $|\mathbf{a}| = \sqrt{4^2 + 3^2 + (-5)^2} = \sqrt{16 + 9 + 25}$
      $= \sqrt{50}.$
    $|\mathbf{b}| = \sqrt{4 + 16 + 36} = \sqrt{56} = 2\sqrt{14}.$
    $|\mathbf{c}| = \sqrt{1 + 25} = \sqrt{26}.$

  b $\mathbf{a} + \mathbf{b} + \mathbf{c} = (4\mathbf{i} + 3\mathbf{j} - 5\mathbf{k}) + (2\mathbf{i} + 4\mathbf{j} + 6\mathbf{k})$
              $+ (-\mathbf{j} + 5\mathbf{k})$
         $= 6\mathbf{i} + 6\mathbf{j} + 6\mathbf{k}.$

  c $\mathbf{a} - \mathbf{b} = (4\mathbf{i} + 3\mathbf{j} - 5\mathbf{k}) - (2\mathbf{i} + 4\mathbf{j} + 6\mathbf{k})$
      $= 2\mathbf{i} - \mathbf{j} - 11\mathbf{k}$

  d $\mathbf{a} \cdot \mathbf{b} = (4\mathbf{i} + 3\mathbf{j} - 5\mathbf{k})(2\mathbf{i} + 4\mathbf{j} + 6\mathbf{k}) = 8 + 12 - 30$
      $= -10.$

  e $\mathbf{a} \times \mathbf{b} = (4\mathbf{i} + 3\mathbf{j} - 5\mathbf{k}) \times (2\mathbf{i} + 4\mathbf{j} + 6\mathbf{k});$ hence,

$$\begin{vmatrix} \mathbf{i} & \mathbf{j} & \mathbf{k} \\ 4 & 3 & -5 \\ 2 & 4 & 6 \end{vmatrix}$$

$$= (18 + 20)\mathbf{i} + (-10 - 24)\mathbf{j} + (16 - 6)\mathbf{k}$$
$$= 38\mathbf{i} - 34\mathbf{j} + 10\mathbf{k}$$

$$\mathbf{b} \times \mathbf{a} = \begin{vmatrix} \mathbf{i} & \mathbf{j} & \mathbf{k} \\ 2 & 4 & 6 \\ 4 & 3 & -5 \end{vmatrix}$$

$$= (-20 - 18)\mathbf{i} + (24 + 10)\mathbf{j} + (6 - 16)\mathbf{k}$$
$$= -38\mathbf{i} + 34\mathbf{j} - 10\mathbf{k} = -\mathbf{a} \times \mathbf{b}$$

    Notice that $\mathbf{a} \times \mathbf{b} \neq \mathbf{b} \times \mathbf{a}$, but that $\mathbf{a} \times \mathbf{b} = -\mathbf{b} \times \mathbf{a}$.

2. $a_x = 1, a_y = 2, a = \sqrt{1^2 + 2^2} = \sqrt{5}$; thus,
  $a = \mathbf{i}\sqrt{5}\cos\theta + \mathbf{j}\sqrt{5}\sin\theta.$

3. a $9.77 \times 10^9 = 9.8 \times 10^9.$
  b $7.6532 \times 10^{-2} = 7.65 \times 10^{-2} = 7.7 \times 10^{-2}.$
  c $13 \times 10^{-4} = 1.3 \times 10^{-3}.$
  d 1.21
  e $35 \times 10^{-7} = 3.5 \times 10^{-6}.$

Answers

$|\mathbf{a}| = \sqrt{50},\ |\mathbf{b}| = 2\sqrt{14},\ |\mathbf{c}| = \sqrt{26}$	**1** a
$6\mathbf{i} + 6\mathbf{j} + 6\mathbf{k}$	b
$2\mathbf{i} - \mathbf{j} - 11\mathbf{k}$	c
-10	d
$\mathbf{a} \times \mathbf{b} = 38\mathbf{i} - 34\mathbf{j} + 10\mathbf{k}$	e
$\mathbf{b} \times \mathbf{a} = -\mathbf{a} \times \mathbf{b} = -38\mathbf{i} + 34\mathbf{j} - 10\mathbf{k}$	
$a = \mathbf{i}\sqrt{5}\cos\theta + \mathbf{j}\sqrt{5}\sin\theta$	**2**
9.8×10^9	**3** a
7.7×10^{-2}	b
1.3×10^{-3}	c
1.21	d
3.5×10^{-6}	e

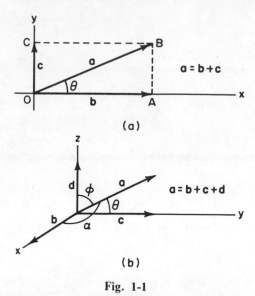

(a)

$$a = b + c$$

(b)

$$a = b + c + d$$

Fig. 1-1

hence **a** is resolved into **b** and **c**. Any one of the four edges of the parallelogram, which are along or parallel to axes *x* and *y*, can be designated as components of **a**. Similarly, in a three dimensional space with axes *x*, *y*, and *z*, a vector can be resolved into *three* components, each placed along or parallel to one of the axes (Fig. 1-1b). For simplifying calculations it is common to use the triangular form with **c** in Fig. 1-1a along *AB* (or with **b** along *CB*). (See also Fig. 1-3.) A similar arrangement can be made for a three dimensional system. A vector may be located at any position in a coordinate system; that is, its tail need not be at the origin *O*, and a vector may have any number of components. It is always understood that a vector is the *sum* of its components. In Fig. 1-2, **f** is the sum of components **a**, **b**, **c**, **d**, and **e**.

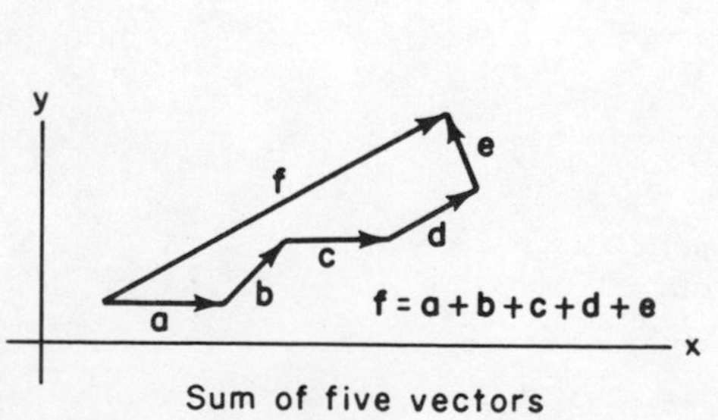

direction of $\mathbf{a} \times \mathbf{b} = -\mathbf{b} \times \mathbf{a}$ and the *magnitude* of $\mathbf{a} \times \mathbf{b}$ is $|\mathbf{a} \times \mathbf{b}| = ab \sin \theta$, where θ is the angle between **a** and **b**.

In terms of the components a_x, a_y, and a_z, $\mathbf{a} \times \mathbf{b} = (a_x\mathbf{i} + a_y\mathbf{j} + a_z\mathbf{k}) \times (b_x\mathbf{i} + b_y\mathbf{j} + b_z\mathbf{k})$. This vector product can be calculated using the following determinant; thus,

$$\mathbf{a} \times \mathbf{b} = \begin{vmatrix} \mathbf{i} & \mathbf{j} & \mathbf{k} \\ a_x & a_y & a_z \\ b_x & b_y & b_z \end{vmatrix}$$

$$= (a_y b_z - a_z b_y)\mathbf{i} + (a_z b_x - a_x b_z)\mathbf{j} + (a_x b_y - a_y b_x)\mathbf{k}$$

This result can be used without calculating the determinant. The cross product holds only for three-dimensional space. Also, $\mathbf{i} \times \mathbf{i} = \mathbf{j} \times \mathbf{j} = \mathbf{k} \times \mathbf{k} = 0$, but $\mathbf{i} \times \mathbf{j} = \mathbf{k}$, $\mathbf{j} \times \mathbf{k} = \mathbf{i}$, and $\mathbf{k} \times \mathbf{i} = \mathbf{j}$.

Sigma Notation.   The Greek letter, uppercase sigma Σ signifies *summation*; thus, $\sum_{i=1}^{n} A_i = A_1 + A_2 + A_3 + \cdots + A_n$. The briefer form ΣA is sometimes used.

Scientific Notation.   In the decimal system, every number can be expressed as a power of 10; that is, $5 = 5 \times 10^0$, $50 = 5 \times 10^1$, $500 = 5 \times 10^2$, etc.; similarly, $0.5 = 5 \times 10^{-1}$, $0.05 = 5 \times 10^{-2}$, $0.005 = 5 \times 10^{-3}$, etc. Thus, $52 = 5.2 \times 10^1$, $436 = 4.36 \times 10^2$, $2594 = 2.594 \times 10^3$, and $0.0365 = 3.65 \times 10^{-2}$, $0.00493 = 4.93 \times 10^{-3}$, etc.

$$f = a + b + c + d + e$$

Sum of five vectors

Fig. 1-2

2

KINEMATICS

1. A car travels 100 ft in 50 sec, turns around in a negligible amount of time, and returns in 25 sec to its starting point.
   a   What is the car's average velocity during the first 50 sec?
   b   What is the car's average velocity during the entire trip?
   c   What is the car's average speed?

2. a   In Fig. 2-1, what is the initial velocity?
   b   What is the velocity at $t = 4$ sec?
   c   What is the average velocity after 8 sec?

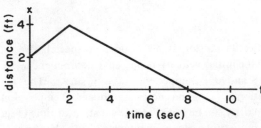

Fig. 2-1

3. A rock is dropped into a well and hits the water 3 sec after being released.
   a   How deep is the well?
   b   If the rock had been dropped at an initial velocity of 200 ft/sec, how long would it have taken to hit the water?
   c   Under the conditions in (b), how fast is the rock traveling when it hits the water?

4. A car accelerates uniformly (a = constant) from 10 m/sec to 60 m/sec in 5 sec.
   a   What is the car's acceleration?
   b   What is the car's velocity after 3 sec?
   c   How far has the car gone after 3 sec?

5. A ball on the end of a string whirls in a circle at 60 cm/sec. The radius of the circle is 30 cm.
   a   What is the period of revolution of the ball?
   b   What is the ball's centripetal acceleration?

1 a
  b
  c

2 a
  b
  c

3 a
  b
  c

4 a
  b
  c

5 a
  b

BASIC FACTS

Kinematics is the study of the motion of bodies, and includes the study of **position, velocity,** and **acceleration** of a particle at every instant of time. Problems in kinematics usually require solutions for the position, velocity, or the time necessary for a particle to attain a certain position or velocity as it follows its path. The position of a particle in three dimensions is given by the vector **r** which usually extends from the origin to a point on the particle's path; thus,

$$\mathbf{r} = x\mathbf{i} + y\mathbf{j} + z\mathbf{k}, \qquad (2\text{-}1)$$

where x, y, and z are the Cartesian coordinates of the point at the arrowhead of the vector **r**, and **i, j,** and **k** are the *unit vectors* in the **x, y,** and **z** directions.

The **average velocity** $\bar{\mathbf{v}}$ of a particle is,

$$\bar{\mathbf{v}} = \frac{\Delta \mathbf{r}}{\Delta t}, \qquad (2\text{-}2)$$

where $\Delta \mathbf{r}$ is the *displacement* of the particle from the starting point of the motion to the final point, and Δt is the total time required for this displacement. Since $\Delta \mathbf{r}$ is a vector, $\bar{\mathbf{v}}$ is a vector. The average speed is,

$$speed_{\text{av}} = \frac{distance\ traveled}{total\ time}. \qquad (2\text{-}3)$$

The total distance may not be the same as $\Delta \mathbf{r}$; for example, when a particle returns to its starting point, $\Delta \mathbf{r} = 0$, but the distance traveled is not zero.

The instantaneous velocity v is,

$$\mathbf{v} = \lim_{\Delta t \to 0} \frac{\Delta \mathbf{r}}{\Delta t} = \frac{d\mathbf{r}}{dt}, \qquad (2\text{-}4)$$

where $d\mathbf{r}/dt$ is the *limit* as Δt approaches zero. In the calculus, $d\mathbf{r}/dt$ is called the *derivative* of **r** with respect to t. In a three-dimensional system with coordinates x, y, and z, $d\mathbf{r}/dt$ can be written,

$$\mathbf{v} = \frac{d\mathbf{r}}{dt} = \frac{dx}{dt}\mathbf{i} + \frac{dy}{dt}\mathbf{j} + \frac{dz}{dt}\mathbf{k}. \qquad (2\text{-}4a)$$

(Continued on page 8)

ADDITIONAL INFORMATION

Since there are three physical dimensions, every vector equation can be expressed as three equations, one for each of the three coordinate axes x, y, and z. The velocity equation Eq. 2-7 then becomes,

$$v_x = v_{0x} + a_x t \qquad (2\text{-}7a)$$
$$v_y = v_{0y} + a_y t$$
$$v_z = v_{0z} + a_z t$$

The position equation Eq. 2-8 breaks up into,

$$x = x_0 + v_{0x}t + \frac{1}{2}a_x t^2 \qquad (2\text{-}8a)$$

$$y = y_0 + v_{0y}t + \frac{1}{2}a_y t^2$$

$$z = z_0 + v_{0z}t + \frac{1}{2}a_z t^2$$

Similarly, Eq. 2-9 can be expressed as,

$$v_x^2 = v_{0x}^2 + 2a_x(x - x_0) \qquad (2\text{-}9a)$$
$$v_y^2 = v_{0y}^2 + 2a_y(y - y_0)$$
$$v_z^2 = v_{0z}^2 + 2a_z(z - z_0)$$

Projectile motion is an important special case in which the initial velocity v_0 has an x-component $v_{0x} = v_0 \cos\theta$ and a y-component $v_{0y} = v_0 \sin\theta$. The acceleration $a = -g\mathbf{j}$ has no x-component. The position and velocity can be found at every instant of time (Fig. 2-2). For a projectile, special forms of the above equations are,

$$v_x = v_{0x}, \qquad (2\text{-}7b)$$
$$v_y = v_{0y} - gt, \qquad (2\text{-}7c)$$
$$x = x_0 + v_{0x}t, \qquad (2\text{-}8b)$$
$$y = y_0 + v_{0y}t - \frac{1}{2}gt^2, \qquad (2\text{-}8c)$$

and
$$v_y^2 = v_{0y}^2 - 2gy, \qquad (2\text{-}9b)$$

where $g = 980\ \text{cm/sec}^2 = 9.8\ \text{m/sec}^2$, or $g = 32\ \text{ft/sec}^2$. (Usually the origin is taken as $x_0 = 0$ and $y_0 = 0$.)

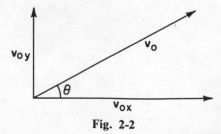

Fig. 2-2

(Continued on page 8)

EXPLANATIONS

1. a $\quad v = \dfrac{\Delta r}{\Delta t} = \dfrac{100 \text{ ft}}{50 \text{ sec}} = 2 \text{ ft/sec.}$

   b $\quad \bar{v} = \dfrac{\Delta r}{\Delta t} = 0,$ because $\Delta r = 0.$

   c $\quad \text{Speed}_{av} = \dfrac{200 \text{ ft}}{75 \text{ sec}} = 2.67 \text{ ft/sec.}$

2. a $\quad v = \dfrac{dx}{dt} = \dfrac{2 \text{ ft}}{2 \text{ sec}} = 1 \text{ ft/sec,}$ since the slope of the line for $0 \leq t < 2$ is 1.

   b $\quad v = \dfrac{\Delta x}{\Delta t} = -\dfrac{2 \text{ ft}}{3 \text{ sec}} = -0.67 \text{ ft/sec;}$ the slope of the line for $t > 2$ is $-\dfrac{2}{3} = -0.67.$

   c $\quad \bar{v} = \dfrac{\Delta x}{\Delta t} = -\dfrac{2 \text{ ft}}{8 \text{ sec}} = -0.25 \text{ ft/sec.}$ This is the slope on the x versus t curve of the line from the origin to the point $x = 2, t = 8.$

3. a $\quad y = -\dfrac{1}{2} gt^2 = -\dfrac{1}{2} (32 \text{ ft/sec}^2)(3 \text{ sec})^2 = -144 \text{ ft.}$

   b $\quad y = v_{0y}t - \dfrac{1}{2} gt^2 = -144 \text{ ft,}$ so that $-144 \text{ ft} = (-200$ ft/sec$)t - \dfrac{1}{2} (32 \text{ ft/sec}^2) t^2$; thus, $t^2 + \dfrac{200}{16} t - \dfrac{144}{16} = 0$ and $t = -\dfrac{200}{32} \pm \dfrac{1}{2} \sqrt{\left(\dfrac{200}{16}\right)^2 + 4 \left(\dfrac{144}{16}\right)}$; using the positive root, $t = 0.69 \text{ sec.}$

   c $\quad v = v_0 - gt = -222 \text{ ft/sec.}$

4. a $\quad v = v_0 + at, \; a = \dfrac{v - v_0}{t} = \dfrac{(60 - 10)\text{m/sec}}{5 \text{ sec}} = 10$ m/sec^2.

   b $\quad v = v_0 + at = 10 \text{ m/sec} + (10 \text{ m/sec}^2)(3 \text{ sec}) = 40$ m/sec.

   c $\quad x = v_0 t + \dfrac{1}{2} at^2 = (10 \text{ m/sec})(3 \text{ sec}) + (10 \text{ m/sec}^2)$ $(3 \text{ sec})^2 = 75 \text{ m.}$

5. a $\quad v = \dfrac{2\pi R}{T}$ and $T = \dfrac{2\pi R}{v} = \dfrac{2\pi (30 \text{ cm})}{60 \text{ cm/sec}}$

   $\qquad = \pi \text{ sec} = 3.14 \text{ sec.}$

   b $\quad a_c = \dfrac{v^2}{R} = \dfrac{(60 \text{ cm/sec})^2}{30 \text{ cm}} = 120 \text{ cm/sec}^2.$

Answers

2 ft/sec	1 a
0	b
2.67 ft/sec	c
1 ft/sec	2 a
−0.67 ft/sec	b
−0.25 ft/sec	c
−144 ft	3 a
0.69 sec	b
−222 ft/sec	c
10 m/sec^2	4 a
40 m/sec	b
75 m	c
3.14 sec	5 a
120 cm/sec^2	b

The **average acceleration** $\bar{a}$ of a particle,

$$\bar{a} = \frac{\Delta \mathbf{v}}{\Delta t} = \frac{\mathbf{v}_f - \mathbf{v}_i}{t_f - t_i}, \qquad (2\text{-}5)$$

where the subscripts f and i stand for "final" and "initial," respectively. The instantaneous acceleration a, the rate of change of the particle's position, is,

$$a = \lim_{\Delta t \to 0} \frac{\Delta \mathbf{v}}{\Delta t} = \frac{d\mathbf{v}}{dt}$$

$$= \frac{dv_x}{dt} \mathbf{i} + \frac{dv_y}{dt} \mathbf{j} + \frac{dv_z}{dt} \mathbf{k} \qquad (2\text{-}6)$$

$$= a_x \mathbf{i} + a_y \mathbf{j} + a_z \mathbf{k}.$$

If a is constant, the velocity $\mathbf{v}$ is

$$\mathbf{v} = \mathbf{v}_0 + \mathbf{a}t, \qquad (2\text{-}7)$$

where $\mathbf{v}_0$ is the velocity of the particle at $t = 0$. If a is constant, the equation for the position r as a function of time is,

$$r = r_0 + \mathbf{v}_0 t + \frac{1}{2} \mathbf{a}t^2, \qquad (2\text{-}8)$$

where r_0 is the position of the particle at $t = 0$. Also, a very useful equation is,

$$v^2 = v_0^2 + 2a\Delta r. \qquad (2\text{-}9)$$

For uniform circular motion, the particle has a **centripetal acceleration** a_c,

$$a_c = \frac{v^2}{R}, \qquad (2\text{-}10)$$

where R is the radius. This equation holds even if the motion is through only part of a circle. When the period T (the time required for a particle to move around a full circle) is given,

$$v = \frac{\text{circumference}}{\text{period}} = \frac{2\pi R}{T}; \quad (2\text{-}11)$$

thus, $\qquad a_c = \dfrac{4\pi^2 R}{T^2}. \qquad (2\text{-}10a)$

The problem of two-dimensional projectile motion breaks up into two one-dimensional problems. The path of a projectile is shown in Fig. 2-3a. The velocity in the x-direction (Fig. 2-3b) is constant and equal to the slope of the x versus the t curves (Fig. 2-3c). In Fig. 2-3d the y-distance, or height, increases to a maximum; at this time v_y (Fig. 2-3e) is zero. The projectile then falls and v_y becomes negative. The instantaneous slope of y versus t is v_y ($v_y = dy/dt$).

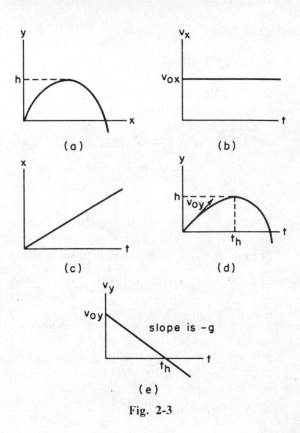

Fig. 2-3

In the case of uniform circular motion, the magnitude v of the vector $\mathbf{v}$ is constant, but the direction of $\mathbf{v}$ is along the tangent to the circle. The acceleration a_c is constant in magnitude, but its direction is always toward the center of the circle (Fig. 2-4).

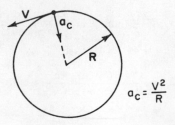

Fig. 2-4

3

1. If a stone weighs 48 lb, what is its mass?

2. If a log is pulled by a car at a constant velocity of 2 ft/sec, what is the net force on the log?

3. In Fig. 3-1, a 300 kg block is pulled along a table top. The coefficient of friction between the table and block is μ = 0.4.
   a Find the normal force of the block on the table.
   b Find the acceleration of the block.

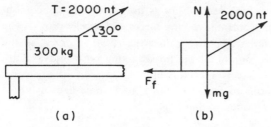

(a)   (b)

Fig. 3-1

4. A 30 gm ball at the end of a string is swung in a verticle circle with a radius of 25 cm. The velocity v = 200 cm/sec. Find the tension in the string: a) at the top of the circle, b) at the bottom of the circle, and c) at the midpoint of the string.

5. A 20 kg box slides down a plane inclined at 30° with the horizontal. The coefficient of friction between the box and the plane is μ = 0.03.
   a Find the normal component of force between the box and the plane.
   b Find the net force, $\Sigma\mathbf{F}$ on the box and the acceleration of the box.
   c Find the velocity of the box as it reaches the end of the plane, if the length of the plane is 2 m and the block starts at rest.

6. An Atwood machine (Fig. 3-2) has mass m_1 = 2 kg and mass m_2 = 4 kg. (Consider the pulley to be frictionless.)
   a What force must a person use to keep m_2 from moving?
   b If m_2 is released, what will its acceleration be?
   c What is the tension in the string?

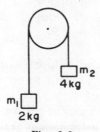

Fig. 3-2

1	
2	
3	a
	b
4	a
	b
	c
5	a
	b
	c
6	a
	b
	c

BASIC FACTS

Newton's Laws

1. A body is at *equilibrium* when its acceleration a is zero. **Newton's first law** states that, at equilibrium the sum of the forces on a body is zero. Thus,

$$\sum_{i=0}^{n} \mathbf{F}_i = 0 \qquad (3\text{-}1)$$

2. **Newton's second law** states that the sum of the forces on a body, called the **net force**, is proportional to the acceleration. Thus, the net force,

$$\sum_{i=0}^{n} \mathbf{F}_i = ma, \qquad (3\text{-}2)$$

where $\mathbf{a}$ is the acceleration of the body, and m is the **mass** of the body.

3. **Newton's third law** of motion states that whenever two bodies interact (exert forces on one another), force $\mathbf{F}_{12}$ on body 1 caused by body 2, is equal in magnitude to but opposite in direction from force $\mathbf{F}_{21}$ on body 2 caused by body 1; thus $\mathbf{F}_{12} = -\mathbf{F}_{21}$.

Mass is often confused with weight. Weight has the units of force; therefore, the weight w of an object is $w = mg$, where m is the mass in gm, kg, or slugs, and g is the **acceleration of gravity** in ft/sec², m/sec², or cm/sec². If the weight is given, it must be divided by g to find m.

An essential technique in solving problems using Newton's laws is the use of **free body diagrams** (described in this chapter under Additional Information). Each force acting on a body must be represented by the force vector $\mathbf{F}$. The resultant of all these vectors is $\Sigma \mathbf{F}$. The vector $\mathbf{a}$ is parallel to $\Sigma \mathbf{F}$; thus,

$$a = \frac{\Sigma \mathbf{F}}{m}. \qquad (3\text{-}3)$$

(Continued on page 12)

ADDITIONAL INFORMATION

Newton's first law is an important special case of Newton's second law. When the sum of the forces exerted on an object is zero, the object moves at a constant velocity (not necessarily zero velocity). If the velocity of the body changes; that is, if it speeds up or slows down or if the direction of motion changes, then the sum of forces is not zero.

The concept of **inertial reference systems** is defined from Newton's first law. A reference system is a coordinate system which may be moving and/or accelerating. The velocity and acceleration of an object can be measured relative to any reference system. An inertial frame is one in which Newton's first and second laws hold for a body. Inertial frames move at constant velocities (including zero velocity). Examples of noninertial frames will be mentioned in chapter 8.

Newton's second law is useful for any nonequilibrium problem involving forces and accelerations. The **net force** $\Sigma \mathbf{F}$, also called the *resultant force*, is the vector sum of all the forces acting on a body. Since $\Sigma \mathbf{F} = ma$ is a vector equation, it can be expressed in terms of the x, y, and z coordinate axes by three corresponding equations,

$$\Sigma F_x = ma_x \qquad (3\text{-}4)$$
$$\Sigma F_y = ma_y$$
$$\Sigma F_z = ma_z$$

To apply Newton's third law correctly, the pair of forces $\mathbf{F}_{12}$ and $\mathbf{F}_{21}$ called "action" and "reaction," respectively, must be carefully identified. Referring to Fig. 3-3, if a man pulls a rope with the force $\mathbf{F}_{12}$,

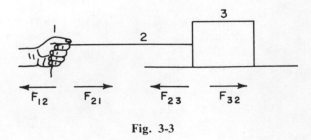

Fig. 3-3

the rope pulls the man with force $\mathbf{F}_{21}$; furthermore, $\mathbf{F}_{12} = -\mathbf{F}_{21}$. Similarly forces $\mathbf{F}_{23}$ and $\mathbf{F}_{32}$ form an

(Continued on page 12)

EXPLANATIONS

1. $m = \dfrac{W}{g} = \dfrac{48 \text{ lb}}{32 \text{ ft/sec}^2} = 1.5$ slug.

2. From Newton's first law, when $a = 0$, $F = 0$.

3. a   The free body diagram is shown in Fig. 3-1b.
$\Sigma F_y = (2000 \text{ nt}) \sin 30° + N - (300 \text{ kg})(9.8 \text{ m/sec}^2) = 0$; thus $N = 1940$ nt.

   b $\Sigma F_x = (2000) \cos 30° - F_f = ma$, where $F_f = \mu N = (0.4)(1940 \text{ nt}) = 776$ nt, and $a = \dfrac{946 \text{ nt}}{300 \text{ kg}} = 3.15$ m/sec^2.

4. a $\Sigma F_y = -T - mg = ma_c = -\dfrac{mv^2}{R}$,

   $T = ma_c - mg = 30 \text{ gm} \dfrac{(200 \text{ cm/sec}^2)}{25 \text{ cm}}$

   $- (30 \text{ gm})(980 \text{ cm/sec}^2) = 1.86 \times 10^4$ dy.

   b $\Sigma F_y = T - mg = mv^2/R$, so that $T = mg + mv^2/R$
   $= (4.8 + 2.94) \times 10^4 \text{ dy} = 7.74 \times 10^7$ dy.

   c $\Sigma F_x = T = ma_c$ and $T = 4.8 \times 10^4$ dy.

5. a $\Sigma F_y = N - mg \cos 30° = 0$; $N = (20 \text{ kg})(9.8 \text{ m/sec}^2)(0.866) = 170$ nt.

   b   Thus $F_f = \mu N = (0.03)(170 \text{ nt}) = 5.1$ nt;

   $\Sigma F_x = mg \sin 30° - F_f = ma = (20 \text{ kg})(9.8 \text{ m/sec}^2) \dfrac{1}{2}$

   $- 5.1 \text{ nt} = 92.9 \text{ nt}; a = \dfrac{\Sigma F_x}{m} = \dfrac{92.2 \text{ nt}}{20 \text{ kg}}$

   $= 4.65$ m/sec^2.

   c $v_f^2 = v_0^2 + 2a\Delta x = 0 + 2(4.65 \text{ m/sec}^2)(2\text{m}) = 18.6$ m^2/sec^2; hence, $v_f = \sqrt{18.6 \text{ m}^2/\text{sec}^2} = 13.6$ m/sec.

6. a T is the same for both strings. For m_2, $\Sigma F = T - m_2 g = 0$; thus $T = m_2 g = (4 \text{ kg})(9.8 \text{ m/sec}^2) = 39.2$ nt; for m_2, $\Sigma F = T - F_{ext} - m_1 g = 0$. Thus, $F_{ext} = 39.2 - (2 \text{ kg})(9.8 \text{ m/sec}^2) = 19.6$ nt.

   b   For m_2, $\Sigma F = T - m_2 g = m_2 a_2 = -m_1 a_1$; for m_1, $\Sigma F = T - m_1 g = m_1 a_1 (a_2 = -a_1$ because both blocks accelerate at the same rate, but in opposite directions.) Subtracting the above equations, $(m_2 - m_1)g = (m_1 + m_2)a_1$; thus, $a_1 = \dfrac{m_2 - m_1}{m_1 + m_2} g = \dfrac{4 - 2}{4 + 2}$ kg$(9.8$ m/sec$^2) = 3.27$ m/sec^2.

   c $T - m_1 g = m_1 a_1$; thus $T = m_1(g + a_1) = 2$ kg$(9.8$ m/sec$^2 + 3.27$ m/sec$^2) = 26.1$ nt.

Answers

1.5 slug	1
0	2
1940 nt	3 a
3.15 m/sec^2	b
1.86 × 10^4 dy	4 a
7.74 × 10^7 dy	b
4.8 × 10^4 dy	c
170 nt	5 a
92.9 nt, 4.65 m/sec^2	b
13.6 m/sec	c
19.6 nt	6 a
3.27 m/sec^2	b
26.1 nt	c

The surface of a body sliding on a surface exerts a frictional force $\mathbf{F}_f = \mu\mathbf{N}$, where μ is the **coefficient of sliding friction** and $\mathbf{N}$ is the normal (perpendicular) force between the surfaces. If the μ for two surfaces is given, the frictional force between them can be calculated by finding the normal force $\mathbf{N}$ exerted by one body on the other. The frictional force always opposes motion and must be included in a free body diagram.

Some of the most common physics problems require the use of the second law for one or more of the following:
1. External forces, such as tension in a rope or an external push, required to accelerate an object.
2. The acceleration of a body subject to one or more forces.
3. Any problem in kinematics (chapter 2) which requires first finding the acceleration of a body.

The units of force are 1 dyne = 1 gm cm/sec^2, 1 newton = 1 kg m/sec^2, or 1 pound = 1 slug ft/sec^2. They are abbreviated as dy, nt, and lb, respectively.

action-reaction pair; that is, $\mathbf{F}_{23} = -\mathbf{F}_{32}$, but forces $\mathbf{F}_{13}$ and $\mathbf{F}_{23}$ do not form an action-reaction pair. They are equal only if the rope is at equilibrium; in that case, $\boldsymbol{a} = 0$ for the rope and Newton's first law, $\Sigma\mathbf{F}_i = \mathbf{F}_{12} - \mathbf{F}_{21} = 0$ applies.

Example. Find the acceleration of the cart m_1 in Fig. 3-3a pulled by the force m_2g. The weights of the string and pulley are negligible. The first step is to draw free body diagrams of carts m_1 and m_2. Using Fig. 3-3c, the net force on m_2 is,

$$\Sigma\mathbf{F}_2 = \mathbf{T}_2 - m_2\mathbf{g} = m_2\boldsymbol{a},$$

where $\mathbf{T}$ is the *tension*. Notice that the forces are in opposite directions, hence the minus sign. Since the masses of the pulley and string are ignored, $\mathbf{T}_1 = \mathbf{T}_2 = \mathbf{T}$. Also $\boldsymbol{a}_1 = \boldsymbol{a}_2 = \boldsymbol{a}$, since, if m_1 moves to the right (positively), m_2 moves down (negatively) just as far. For the cart,

and
$$\Sigma\mathbf{F}_{1y} = \mathbf{N} - m_1\mathbf{g} = 0$$
$$\Sigma\mathbf{F}_{1x} = \mathbf{T} = m_1\boldsymbol{a}.$$

Hence, normal force $\mathbf{N} = m_1\mathbf{g}$ and,

thus,
$$\mathbf{T} - m_2\mathbf{g} = m_2\boldsymbol{a}$$
$$\mathbf{T} = m_1\boldsymbol{a}.$$

Solving for $\boldsymbol{a}$, then,

$$\boldsymbol{a} = \frac{m_2\mathbf{g}}{m_1 - m_2}.$$

In all problems involving Newton's laws, draw a **free body diagram** using the following rules:
1. Establish an inertial reference frame; that is, a system of coordinate axes with an origin, for each direction (x, y, z, . . .) necessary to represent a body's environment.
2. Draw only the body, not specific details, such as a surface supporting it, a string suspending it, or a wall pushing it.
3. Represent all the forces acting on the body by arrows; for example, a string holding the body up is represented by an arrow away from the body, in the same direction as the string; the string itself is not shown. Always include an arrow representing the weight of the body.
4. Add all the forces using vector addition (chapter 1). It is often simpler to find $\Sigma\mathbf{F}_x$ and $\Sigma\mathbf{F}_y$.

4

WORK AND ENERGY

1. A man pushes against a stationary wall with a force of 200 nt. Is he doing any work?

2. A man pushes a 20 kg block at a constant velocity up a frictionless inclined plane a distance of 40 m. The plane is inclined at an angle of 30°. What is the work done by the man?

3. The brakes of a 2000 lb car slip, causing the car to roll down a hill a total distance of 50 ft. The frictional force is 100 lb and the point at which the brake slips is 10 ft above the bottom of the hill.
   a What is the change in potential energy of the car?
   b What is the change in kinetic energy of the car?

4. A 3 kg mass is placed on a vertical spring and the spring compresses to 3 cm.
   a What is the spring constant?
   b What is the potential energy of the mass relative to its energy before compression?

5. An 8 lb rock falls from rest and drops 100 ft and stops falling when it hits the ground.
   a What is the velocity of the rock just before it hits the ground?
   b How much energy is dissipated during the collision?

6. A boy walking at the rate of 2 ft/sec pulls a toy by a string with a force of 4 lb. The string makes an angle of 60° with the horizontal.
   a What is the boy's power output?
   b How much work will the boy do in 3 minutes?

7. An 8 gm mass on the end of a weightless spring is pulled 3 cm from equilibrium and is then released. If the spring constant is 20 dy/cm, when the mass passes through the equilibrium point again, what is its velocity?

8. A 20 kg box is pulled at a constant velocity up an inclined plane a distance of 50 m, which is 30 m above the horizontal. The coefficient of friction between the box and the incline is $\mu = 0.3$.
   a Find the normal force between the box and the incline.
   b Find the work done in dragging the box up the plane.
   c What is the energy dissipated during this process?

9. Every minute, 10^2 m^3 of water drop 300 m from a dam to a power generator. What is the optimum power output possible for the generator, if the density of water is 10^3 kg/m^3?

1	
2	
3	a
	b
4	a
	b
5	a
	b
6	a
	b
7	
8	a
	b
	c
9	

BASIC FACTS

When a force F moves a particle a distance Δs, it does an amount of work W; in terms of vector notation,

$$\mathbf{W} = \mathbf{F} \cdot \Delta \mathbf{s} = F \Delta s \cos \theta; \quad (4\text{-}1)$$

or,

$$\mathbf{W} = \int_a^b F \cos \theta \, ds.* \quad (4\text{-}2)$$

The **kinetic energy** K of a body is energy due solely to its velocity; thus,

$$K = \frac{1}{2} m v^2.$$

The **potential energy** U of a body is the part of its energy due solely to its position. Two important examples are (1) gravitational potential energy,

$$U_{\text{grav}} = mgz, \quad (4\text{-}3)$$

where z is the distance above the earth's surface, and (2) elastic, or spring, potential energy,

$$U_{\text{spring}} = \frac{1}{2} k x^2, \quad (4\text{-}4)$$

where k is the spring constant and x is the displacement of the end of the spring from its equilibrium position.

A force is *conservative* if the work done by it in moving a body around a *closed* path back to the starting point is zero. Otherwise, the force is *nonconservative*. If a nonconservative force $\mathbf{F}_n$ is present, then,

$$\mathbf{W}_n = \mathbf{F}_n \cdot \Delta \mathbf{s} = \Delta K + \Delta U.$$
$$(4\text{-}5)$$

*The calculus symbol $\int_a^b$ is called the integral sign and a and b are the lower and upper limits.

(Continued on page 16)

ADDITIONAL INFORMATION

One of the most important uses of the concepts of work and energy is in solving certain problems in dynamics when the forces are conservative. For example, if a weight slides down a frictionless slope, its final velocity can be easily calculated by using the equation for the conservation of energy. To solve this problem using Newton's second law would be impossible, unless the shape of the slope is known, and even then, the problem would be difficult for all but the simplest shapes.

The work performed on a body by the resultant force $\mathbf{F}_{\text{res}} = \Sigma F$ is equal to the change in its kinetic energy ΔK. Thus $\mathbf{F}_{\text{res}} \cdot \Delta \mathbf{s} = F_{\text{res}} \Delta s \cos \theta = \Delta K$; also, $\Delta K = \int \mathbf{F}_{\text{res}} \cdot d\mathbf{s}$ and $\mathbf{F}_{\text{res}} = \mathbf{F}_n + \mathbf{F}_c$, where $\mathbf{F}_n$ is the *nonconservative force* and $\mathbf{F}_c$ is the *conservative force*. The conservative force is the *rate of change* of the potential; therefore, $\mathbf{F}_c \cdot \Delta \mathbf{s} = -\Delta U$ and $\mathbf{F}_n \cdot \Delta \mathbf{s} = \Delta K + \Delta U$. Using the calculus, $-\Delta U = \int \mathbf{F}_c \cdot d\mathbf{s}$ and $\Delta K + \Delta U = \int \mathbf{F}_n \cdot d\mathbf{s}$.

Example 1. Consider a 12 nt force which pushes a 0.5 kg mass 20 m up a 30° inclined plane. If the frictional force is 4 nt, what is the change in kinetic energy? By the first method, $F_n = (12 - 4)$ nt $= 8$ nt, $F_c = -mg \sin 30° = -(0.5 \text{ kg})(9.8 \text{ m/sec}^2)(1/2) = -2.45$ nt, and $F_{\text{res}} = F_n + F_c = (8 - 2.45)$ nt $= 5.55$ nt; thus, $\Delta K = F_{\text{res}} \Delta s = (5.55 \text{ nt})(20 \text{ m}) = 111$ nt m. By the second method, $\Delta U = mg \Delta z = 49$ nt m, since $mg = 4.9$ nt and $\Delta z = (20 \sin 30°)$ m $= (20 \times 1/2)$ m $= 10$ m. Thus, $\Delta K + \Delta U = F_n \Delta s = \Delta K + 49$ nt m $= (8 \text{ nt})(20 \text{ m}) = 160$ nt m and $\Delta K = (160 - 49)$ nt m $= 111$, as before.

Since the potential energy is a function of position only, the difference in potential energy ΔU between two points can be found merely by subtracting. Thus, $\Delta U = U_2 - U_1$ and ΔU is independent of the path taken by the body in moving from point 1 to point 2. For a spring, $U = 1/2 k x^2$ and $F = -dU/dx = -kx$, where x is displacement from the equilibrium position of the spring and k is the spring constant in nt/m, dy/cm, or lb/in.

The work done by frictional forces is *always* negative. Therefore, this work will not be zero unless friction is zero everywhere. The work is $\Delta E = \mathbf{F}_f \cdot \Delta \mathbf{s}$, also $\Delta E = \int \mathbf{F}_f \, ds$, and it appears as heat energy. For the special case of sliding friction, $\Delta E = -\mu \mathbf{N} \cdot \Delta \mathbf{s}$.

(Continued on page 16)

EXPLANATIONS

1. No. In this case no displacement of the wall occurs, thus no work is performed.

2. $W = \Delta K + \Delta U$; ΔK is zero, thus $W = mg\Delta z = (20 \text{ kg})(9.8 \text{ m/sec}^2)(40 \text{ m})(\sin 30°) = 392 \text{ nt m} = 392 \text{ joule}$.

3. a $\Delta U = mg\Delta z = (2000 \text{ lb})(-10 \text{ ft}) = -2 \times 10^4 \text{ ft lb}$.
   b Using Eq. 4-5, $\mathbf{F}_n \cdot \Delta \mathbf{s} = (-100 \text{ lb})(500 \text{ ft}) = -5 \times 10^3$ ft lb, $\Delta K = \mathbf{F}_n \Delta s - \Delta U = 1.5 \times 10^4 \text{ ft lb}$.

4. a $F = -kx$, thus $k = -\dfrac{F}{x} = -\dfrac{(3 \text{ kg})(9.8 \text{ m/sec}^2)}{(-0.03 \text{ m})}$
   $= 980 \text{ nt/m}$.
   The fact that the spring exerts a force equal to the gravitational force when the mass is at equilibrium has been used; this follows from Newton's second law (chapter 3). Notice that k is always positive.
   b $\Delta U = \dfrac{1}{2} k\Delta x^2 + mg\Delta z = \dfrac{1}{2} (980 \text{ nt/m})(0.03 \text{ m})^2 +$
   $(2.94 \text{ dy})(-0.03 \text{ m}) = 0.3528 \text{ nt m} = 0.3528 \text{ joule}$.

5. a Energy is conserved until the rock hits the ground so that
   $\Delta K = -\Delta U = -mg\Delta z = -(8 \text{ lb})(-100 \text{ ft}) = 800 \text{ ft lb}$,
   $m = \dfrac{8 \text{ lb}}{32 \text{ ft/sec}^2} = 0.25$ slug, and $v = \sqrt{\dfrac{2\Delta K}{m}} =$
   $\sqrt{\dfrac{2(800 \text{ ft lb})}{0.25 \text{ slug}}} = 80 \text{ ft/sec}$.
   b The energy is entirely kinetic and all of it is dissipated in the form of heat; thus, $\Delta E_{\text{dissip}} = 800 \text{ ft lb}$.

6. a $P = Fv\cos 60° = (4 \text{ lb})(2 \text{ ft/sec}) \dfrac{1}{2} = 4 \text{ ft lb/sec}$.
   b $W = Pt = (4 \text{ ft lb/sec})(3 \times 60 \text{ sec}) = 720 \text{ ft lb/sec}$.

7. The potential energy before the mass is released is
   $U = \dfrac{1}{2} kx^2 = \dfrac{1}{2} (20 \text{ dy/cm})(3 \text{ cm})^2 = 90 \text{ dy cm} = 90 \text{ erg}$.
   Thus $K = \dfrac{1}{2} mv^2 = 90 \text{ dy cm} = \dfrac{1}{2} (8 \text{ gm})v^2$; $v = \sqrt{\dfrac{90 \text{ dy cm}}{4 \text{ gm}}}$
   $= 4.74 \text{ cm/sec}$.

8. a $N = mg\cos\theta$. Since the incline forms a 3-4-5 right triangle, $\cos\theta = 4/5$, $N = (20 \text{ kg})(9.8 \text{ m/sec}^2)\left(\dfrac{4}{5}\right) = 156.8 \text{ nt}$.
   b $W = \mathbf{F}_n \cdot \Delta \mathbf{s} = \Delta U = mg\Delta z = (20 \text{ kg})(9.8 \text{ m/sec}^2)(30 \text{ m}) = 5880 \text{ joule}$.
   c The energy dissipated is $\Delta E = \mu N\Delta s = (0.3)(156.8 \text{ nt})(50 \text{ m}) = 2352 \text{ nt}$.

9. The optimum power output occurs when the water has zero velocity after passing through the generator. In this case, the kinetic energy of the water is converted entirely into useful power. The kinetic energy before passing through the generator is equal to the change in potential energy change of the water $mg\Delta z$. The power output is therefore,
   $$P = \frac{\Delta E}{\Delta t} = \frac{mg\Delta z}{\Delta t}$$
   $$= \frac{(10^2 \text{ m}^3)}{60 \text{ sec}} (10^3 \text{ kg/m}^3)(9.8 \text{ m/sec}^2)(300 \text{ m}) = 4.9 \times 10^6 \text{ watts}.$$

Answers

No	1
392 joule	2
-2×10^4 ft lb	3 a
1.5×10^4 ft lb	b
980 nt/m	4 a
0.3528 joule	b
80 ft/sec	5 a
800 ft lb	b
4 ft lb/sec	6 a
720 ft lb/sec	b
4.74 cm/sec	7
156.8 nt	8 a
5880 joule	b
2352 nt	c
4.9×10^6 watts	9

If either F or $\cos\theta$ varies along path s,

$$\mathbf{W}_n = \int F_n \cos\theta \, ds = \Delta K + \Delta U, \quad (4\text{-}6a)$$

where in both equations, $\Delta K = K_2 - K_1$ and $\Delta U = U_2 - U_1$. Whenever a body experiences only conservative forces, the total energy E_{tot} is conserved. This means that $E_{tot} = K + U$ is constant and does not depend on either position or time.

Frictional forces dissipate energy and are nonconservative. A common example of a frictional force is **sliding friction** (chapter 3).

Power P is the rate at which work is performed; hence,

$$P = \frac{dW}{dt}. \quad (4\text{-}7)$$

A force F which moves a body at a velocity v expends an amount of power,

$$\mathbf{P} = \mathbf{F} \cdot \mathbf{v}. \quad (4\text{-}8)$$

If the power is a constant, then,

$$W = Pt. \quad (4\text{-}9)$$

If P is *not* constant,

$$W = \int_{t_1}^{t_2} P \, dt. \quad (4\text{-}9a)$$

The **efficiency** of a power transmitter, such as a system of gears and pulleys, is given by,

$$efficiency = \frac{output\ power}{input\ power}.$$

The units of work and energy are 1 joule = 1 nt m, 1 erg = 1 dy cm, or ft lb. The units of power are erg/sec, 1 joule/sec = 1 watt, or 1 ft lb/sec = 1/550 horsepower.

The useful power delivered by an engine is illustrated by the following two examples.

Example 2.   An elevator weighing 6400 lb is pulled up at a constant velocity of 100 ft/sec. How much work is required to lift the elevator 500 ft? The power delivered by the engine lifting the elevator is $\mathbf{P} = \mathbf{F} \cdot \mathbf{v} = (6400 \text{ lb})(100 \text{ ft/sec}) = 6.4 \times 10^5$ ft lb/sec. Since P is constant, $W = Pt$ and,

$$t = \frac{s}{v} = \frac{500 \text{ ft}}{100 \text{ ft/sec}} = 5 \text{ sec}.$$

Thus, $W = Pt = (6.4 \times 10^5 \text{ ft lb/sec})(5 \text{ sec}) = 3.2 \times 10^6$ ft lb.

Example 3.   The elevator in example 2 is accelerated uniformly upward from rest to 100 ft/sec in 10 sec. How much work is done in this time? The power required is not constant in time because the velocity is not constant, therefore the work $W = \int P \, dt$. Now, $P = Fv = Fat$, $F = m(g + a)$, $a = v/t$, and $m = w/g$; thus,

$$m = \frac{w}{g} = \frac{6400 \text{ lb}}{32 \text{ ft/sec}^2} = 200 \text{ slug};$$

therefore, $F = m(g + a)$
$\qquad\quad = (200 \text{ slug})(32 + 10) \text{ ft/sec}^2$
$\qquad\quad = 8.4 \times 10^3$ lb. The acceleration is,

$$a = \frac{v}{t} = \frac{100 \text{ ft/sec}}{10 \text{ sec}} = 10 \text{ ft/sec}^2.$$

Thus, $P = Fat = (8.4 \times 10^3 \text{ lb})(10 \text{ ft/sec}^2)t$
$\qquad\quad = (8.4 \times 10^4 \text{ ft lb/sec}^2)t$ and,

$$W = \int P \, dt = \int_0^{10 \text{ sec}} (8.4 \times 10^4 \text{ ft lb/sec}^2) t \, dt$$

$$= (8.4 \times 10^4 \text{ ft lb/sec}^2) \left. \frac{t^2}{2} \right|_0^{10 \text{ sec}}$$

$$= 4.2 \times 10^6 \text{ ft lb}.$$

5 CONSERVATION OF LINEAR MOMENTUM

SELF-TEST

1. A 5 gm bullet strikes a stationary 2 kg block, which is free to slide on a frictionless surface. The bullet is imbedded in the block, and both the bullet and the block slide at 21 cm/sec. Find the velocity of the bullet before the impact.

2. A cannon on wheels fires a 1 slug cannon ball horizontally at 500 ft/sec. The mass of the cannon is 200 slug. If the cannon was originally at rest, what is its velocity after the ball is fired?

3. A 40 kg cart moving at 2 m/sec makes a head-on collision with a 20 kg cart that is standing still. After the collision, the second cart moves at 1 m/sec.
   a How fast does the first cart move after the collision?
   b How much energy is conserved in the collision?
   c What is the coefficient of restitution of the carts?

4. The coefficient of restitution for collisions between two surfaces of a certain metal is 0.6. Two 3 kg balls, made of these metals, traveling at 2 m/sec collide head-on.

a What will the velocities of the balls be after the collision?
b How much kinetic energy is lost in the collision?

5. The pendulum shown in Fig. 5-1 swings down and strikes the second pendulum.

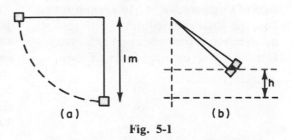

(a)   (b)

Fig. 5-1

Both pendulums stick and swing together, and the mass of both pendulums is 1 kg.
a What is the velocity of the first pendulum just before it hits the second?
b What is the velocity of the combined pendulums just after the collision?
c How high will the combined pendulums rise?
d Assuming that the collision is perfectly elastic, what will happen to the two pendulums after the collision?

1	
2	
3	a
	b
	c
4	a
	b
5	a
	b
	c
	d

17

BASIC FACTS

Linear momentum p is defined as,

$$\mathbf{p} = m\mathbf{v}, \qquad (5\text{-}1)$$

where m is the mass and v is the velocity. If there are two masses m_1 and m_2, the total linear momentum $\mathbf{p}_{tot}$ is,

$$\mathbf{p}_{tot} = m_1\mathbf{v}_1 + m_2\mathbf{v}_2. \qquad (5\text{-}2)$$

All the v's are velocities of the centers of mass of the respective particles. The **Law of Conservation of Momentum** states that the total momentum is constant if no external forces act upon the system. If there is a net external force $\Sigma\mathbf{F}$ then,

$$\Sigma\mathbf{F} = \frac{d\mathbf{p}}{dt}. \qquad (5\text{-}3)$$

A **collision** between particles is *perfectly elastic*, if the total kinetic energy before the collision K_b equals the total kinetic energy after the collision K_a; thus,

$$\tfrac{1}{2}\,m_1 v_{1b}^2 + \tfrac{1}{2}\,m_2 v_{2b}^2 = \tfrac{1}{2}\,m_1 v_{1a}^2 + \tfrac{1}{2}\,m_2 v_{2a}^2, \qquad (5\text{-}4)$$

where the subscripts b and a stand for "before" and "after," respectively. Another equivalent method of describing perfectly elastic collisions is to say that the *magnitude* $|\mathbf{v}_1 - \mathbf{v}_2|$ of the relative velocities of the two particles is the same before and after the collision; thus,

$$|v_{1b} - v_{2b}| = |v_{1a} - v_{2a}|. \qquad (5\text{-}5)$$

If the collision is *inelastic*, then $K_b > K_a$ and $|v_{1b} - v_{2b}| > |v_{1a} - v_{2a}|$; some of the kinetic energy went into deforming or heating the bodies. The extreme case of an inelastic collision is one in which the bodies stick together. This case is called a **perfectly inelastic collision.** Problems in which two bodies interact, such as when two rocks collide or a rifle

(Continued on page 20)

ADDITIONAL INFORMATION

The **center of mass** is the point of a body or system of bodies which moves as if all the mass were concentrated at that point. Thus a stick thrown into the air may turn about, but its center of mass follows a simple parabola. Kinematics, discussed in chapter 2, deals with the motion of the center of mass of bodies.

The momentum of a system of particles is a constant. This means that momentum is conserved in any interaction between the bodies. Here are two examples.

Example 1. A 5 gm bullet in a 3 kg gun is shot. If the gun was originally at rest and the muzzle velocity of the bullet is 300 m/sec, what is the gun's recoil velocity? The momentum p_b of the system is the sum of the momentum p_{b1} of the gun and the momentum p_{b2} of the bullet before it is fired; thus $\mathbf{p}_b = \mathbf{p}_{1b} + \mathbf{p}_{2b} = 0$. Each is zero because $v_1 = 0$ and $v_2 = 0$. After firing,

$$p_{2a} = m_{2a}v_{2a} = (0.005\,\text{kg})(300\,\text{m/sec})$$
$$= 1.5\,\text{kg m/sec},$$

and $\mathbf{p}_{1a} + \mathbf{p}_{2a} = 0$; thus $\mathbf{p}_{1a} = -\mathbf{p}_{2a} = -1.5$ kg m/sec, and $v_{1a} = \dfrac{p_{1a}}{m_1} = \dfrac{-1.5\,\text{kg m/sec}}{3\,\text{kg}} = -0.5\,\text{m/sec}.$

Example 2. A boy is standing on a raft; both are initially at rest. The boy starts walking at 5 ft/sec. If the mass m_1 of the boy is 6 slug, and the mass m_2 of the raft is 2 slugs, find the velocity of the raft. The momenta of the boy $\mathbf{p}_{1b}$ and of the raft $\mathbf{p}_{2b}$ before the boy starts walking are both zero. After the boy starts walking, his momentum is $p_{1a} = m_1 v_{1a} = (6$ slugs$)$ $(5$ ft/sec$) = 30$ slug ft/sec. The momentum of the raft is found by equating the total momentum after the boy starts walking to zero; thus,

$$p_{2a} = -30\,\text{slug ft/sec}$$

and,
$$v_{2a} = \frac{-30\,\text{slug ft/sec}}{2\,\text{slug}} = -15\,\text{ft/sec}.$$

Since the momentum equation $\mathbf{p}_b = \mathbf{p}_a$ is a vector equation, the momentum in each direction must be conserved. Thus, for a collision, $\mathbf{p}_{xb} = \mathbf{p}_{xa}$, $\mathbf{p}_{yb} = \mathbf{p}_{ya}$, and $\mathbf{p}_{zb} = \mathbf{p}_{za}$, where $\mathbf{p}_{xb}$ is the x-component of the total momentum before the interaction, $\mathbf{p}_{xa}$ is the x-

(Continued on page 20)

EXPLANATIONS

1. $p_b = p_{1b} + p_{2b} = m_1 v_{1b} + m_2 v_{2b} = (5 \text{ gm}) v_{1b} + (2000 \text{ gm})$ (0), and $p_b = p_a = p_{1a} + p_{2a} = m_1 v_{1a} + m_2 v_{2a} = (5 \text{ gm}) (20 \text{ cm/sec}) + (2000 \text{ gm}) (20 \text{ cm/sec}) = 4.01 \times 10^4 \text{ gm cm/sec}$. Since $p_b = p_a$, $v_{1b} = \dfrac{p_b}{m_1} = \dfrac{4.01 \times 10^4 \text{ gm cm/sec}}{5 \text{ gm}} = 8.02 \times 10^3 \text{cm/sec}$.

2. $p_b = p_{1b} + p_{2b} = 0 + 0 = 0$ and, since $p_a = p_b$, $p_a = p_{1a} + p_{2a} = m_1 v_{1a} + m_2 v_{2a} = (1 \text{ slug})(500 \text{ ft/sec}) + (200 \text{ slug}) v_{2a} = 0$; thus, $v_{2a} = -2.5 \text{ ft/sec}$.

3. a $p_b = p_{1b} + p_{2b} = m_1 v_{1b} + m_2 v_{2b} = (40 \text{ kg})(2 \text{ m/sec}) + (20 \text{ kg})(0) = 80 \text{ kg m/sec}$ and $p_a = p_{1a} + p_{2a} = (40 \text{ kg})(v_{1a}) + (20 \text{ kg})(1 \text{ m/sec}) = 80$ kg m/sec, since $p_b = p_a$; thus, $v_{1a} = 1.5 \text{ m/sec}$.

   b $K_{1b} = \frac{1}{2} m_1 v_{1b}^2 = \frac{1}{2} (40 \text{ kg})(2 \text{ m/sec})^2 = 80$ joule.

   $K_{2b} = \frac{1}{2} m_2 v_{2b}^2 = \frac{1}{2} (20 \text{ kg})(0)^2 = 0$ joule.

   $K_{1a} = \frac{1}{2} m_1 v_{1a}^2 = \frac{1}{2} (40 \text{ kg})(1.5 \text{ m/sec})^2 = 45$ joule.

   $K_{2a} = \frac{1}{2} m_2 v_{2a}^2 = \frac{1}{2} (20 \text{ kg})(1 \text{ m/sec})^2 = 10$ joule.

Thus $K_b = 80$ joule and $K_a = 55$ joule, so that $K_b > K_a$ and $K_b - K_a = 25$ joule.

   c $e = \dfrac{|v_{1a} - v_{2a}|}{|v_{1b} - v_{2b}|} = \dfrac{1.5 - 1}{2 - 0} = 0.25$; since e is a ratio, it has no units.

4. a $p_b = p_{1b} + p_{2b} = 0$, hence $p_{1b} = -p_{2b}$ and $v_{1b} = -v_{2b}$. Because of the law of conservation of momentum, $p_a = p_b = 0$, thus $v_{1a} = -v_{2a}$. $|v_{1b} - v_{2b}| = |2 - (-2)|$ m/sec $= 4$ m/sec and $\dfrac{|v_{1a} - v_{2a}|}{4 \text{ m/sec}} = 0.6$; hence $|v_{1a} - v_{2a}| = 2.4$ m/sec; thus, $v_{1a} = 1.2 \text{ m/sec}$.

   b $K_b = \frac{1}{2} m_1 v_{1b}^2 + \frac{1}{2} m_2 v_{2b}^2 = \frac{1}{2} (3 \text{ kg})(2 \text{ m/sec})^2 + \frac{1}{2} (3 \text{ kg})(2 \text{ m/sec})^2 = 12$ joule. Similarly, $K_a = 4.32$ joule and $\Delta K = K_b - K_a = 7.68$ joule.

5. a By using the law of conservation of energy, the velocity v_{1b} of the first pendulum can be found just before it strikes the second pendulum. The potential energy of the first pendulum before it is released is $U = m_1 g R$. The kinetic energy is zero at this point because it has not started to swing. The total energy is therefore $E = mgR$. When the pendulum has swung down, the potential energy is zero because the height of the pendulum is zero. The total energy on the first pendulum is conserved before it strikes the second pendulum; thus, $K = \frac{1}{2} m v_{1b}^2 = mgR$ and $v_{1b} = \sqrt{2gR} = \sqrt{2(9.8 \text{ m/sec}^2)(1 \text{ m})} = 4.43 \text{ m/sec}$.

   b $p_b = m v_{1b} = (1 \text{ kg})(4.43 \text{ m/sec}) = 4.43 \text{ kg m/sec}$. $p_a =$

Answers

8.02×10^3 cm/sec	1
-2.5 ft/sec	2
1.5 m/sec	3 a
25 joule	b
0.25	c
1.2 m/sec	4 a
7.68 joule	b
4.43 m/sec	5 a
2.22 m/sec	b
0.25 m	c
See explanation 5d.	d

$2mv_a = p_b$; thus, $v_a = \dfrac{p_b}{2m}$

$= \dfrac{4.43 \text{ kg m/sec}}{2(1 \text{ kg})} = 2.22 \text{ m/sec}$.

   c Again using the law of conservation of energy, the height h to which the combined pendulum rises is,

$$h = \frac{v_a^2}{2g} = \frac{(2.22 \text{ m/sec})^2}{2(9.8 \text{ m/sec}^2)} = 0.25 \text{ m}.$$

   d If the collision is perfectly elastic, $K_b = K_a$, that is, $\frac{1}{2} m v_{1b}^2 = \frac{1}{2} m v_{1a}^2 + \frac{1}{2} m v_{2a}^2$. Therefore, $v_{2a}^2 = v_{1b}^2 - v_{1a}^2 = (v_{1b} + v_{1a})(v_{1b} - v_{1a})$, $m v_{1b} = m v_{1a} + m v_{2a}$, and $v_{2a} = v_{1b} - v_{1a} = v_{1b} + v_{1a}$. The last equality was found by substituting v_{2a} into the above equation. Hence $v_{1a} = -v_{1a}$ and, therefore, $v_{1a} = 0$ and $v_{2a} = v_{1b}$. Thus the first pendulum stops and the second pendulum rises 1 m.

fires a bullet, can usually be handled by the law of conservation of momentum.

The x coordinate of the **center of mass** $\bar{x}$ of a set of n particles is

$$\bar{x} = \frac{\sum\limits_{i=1}^{n} m_i x_i}{\sum\limits_{i=1}^{n} m_i}. \qquad (5\text{-}6)$$

The y and z coordinates of the center of mass $\bar{y}$ and $\bar{z}$ can be similarly defined.

The units of momentum are gm cm/sec, kg m/sec, slug ft/sec.

component of the total momentum after the interaction, and so on.

When two bodies collide elastically, their relative velocity before the collision equals their relative velocity after collision. The relative velocity before the collision is,

$$|v_{2b} - v_{1b}|.$$

The relative velocity after the collision is,

$$|v_{2a} - v_{1a}|.$$

If the collision is completely inelastic,

$$|v_{2a} - v_{1a}| = 0.$$

The ratio of the two relative velocities is called the **coefficient of restitution** e and,

$$e = \frac{|v_{1a} - v_{2a}|}{|v_{1b} - v_{2b}|},$$

where $e = 1$ for an elastic collision and $e = 0$ for a completely inelastic collision. A convenient way to measure e for a substance is to drop a sample on a platform that is firmly attached to the earth. Since the velocity of the earth does not change measurably when the sample bounces, the ratio of the initial to the final velocity of the sample is e.

Example 3.   A tennis ball is dropped from 4 ft onto a cement floor and rises only 3 ft after it hits the floor. What is the coefficient of restitution of the tennis ball on the cement? The velocity v_b of the ball when it hits the floor is found by using the law of conservation of energy $\frac{1}{2} mv_b^2 = mgh$; thus,

$$v_b = \sqrt{2gh} = \sqrt{2g\,(4\text{ ft})}$$

The velocity v_a of the ball just *after* it hits the floor is,

$$v_a = \sqrt{2g\,(3\text{ ft.})}. \text{ Then } e = \frac{\sqrt{2g\,(3\text{ ft})}}{\sqrt{2g\,(4\text{ ft})}}$$

$$= \sqrt{\frac{3}{4}} = 0.866.$$

6   ANGULAR MOMENTUM AND TORQUE

1. A flywheel weighing 64 lb and having a radius of gyration of 0.4 ft rotates at 5 rad/sec.
   a What is the moment of inertia of the flywheel?
   b What is the flywheel's angular momentum?
   c What must be the angular acceleration of the flywheel to accelerate it uniformly to 15 rad/sec in 10 sec?

2. In Fig. 6-1, m_2 is released from rest and falls 250 cm in 1 sec; m_1 = 400 gm, m_2 = 50 gm, and r = 10 cm.
   a Find the angular acceleration a of the wheel.

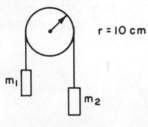

   r = 10 cm

   m_1

   m_2

   Fig. 6-1.

   b Find the moment of inertia for the wheel.
   c Find the radius of gyration for the wheel.

3. A 96 pound boy sits on a swing which is spinning about its vertical axis. When the boy's legs are extended, his radius of gyration k_b (before) is 2 ft, and when he bends his legs, his radius of gyration k_a (after) is 1 ft.
   a If the boy is spinning with an angular frequency of 2π rad/sec (1 rev/sec) with his legs extended outward, how fast would he be spinning if he bends his legs?
   b What is the change in energy of rotation?

4. A flywheel has a moment of inertia of 50 kg m^2 and is initially rotating at 3π rad/sec, but, because of bearing friction, it slows to rest after 20 revolutions.
   a Find the average angular acceleration of the flywheel.
   b How much energy is dissipated by the bearings?

1 a
  b
  c

2 a
  b
  c

3 a
  b

4 a
  b

BASIC FACTS

Angular momentum L is defined as,

$$L = r \times p, \qquad (6\text{-}1)$$

where p is the linear momentum mv and r is the vector from the origin to the mass m (Fig. 6-2).

The distance along a circular path is called the *arc length s* and,

$$s = r\theta, \qquad (6\text{-}2)$$

where r is the radius of the circle and θ is the angle corresponding to the arc length which is measured in *radians*.

The **angular velocity** ω is,

$$\omega = \frac{\Delta\theta}{\Delta t}, \qquad (6\text{-}3)$$

or, $\qquad \omega = \lim_{\Delta t \to 0} \frac{\Delta\theta}{\Delta t} = \frac{d\theta}{dt}, \qquad (6\text{-}3a)$

which is the rate of change of the angle per time. The velocity v of a particle moving in a circle of radius r is,

$$v = r\omega, \qquad (6\text{-}4)$$

where ω is a vector directed along the axis of rotation. The right-hand rule is used to determine the direction of the vector ω. Notice that the fingers in Fig. 6-2 point in the *positive* direction of circular motion and that the thumb points in the direction of ω. If the velocity of a particle moving in a circle of radius r is changing with time; that is, if it is accelerating, then the angular velocity ω is also changing. The change of ω per unit time is called **angular acceleration** α. Thus,

$$\alpha = \frac{\Delta\omega}{\Delta t}, \qquad (6\text{-}5)$$

or, $\qquad \alpha = \lim_{\Delta t \to 0} \frac{\Delta\omega}{\Delta t} = \frac{d\omega}{dt}. \qquad (6\text{-}5a)$

If α is constant, then,

$$\omega = \omega_0 + \alpha t \qquad (6\text{-}6)$$

$$\theta = \omega_0 t + 1/2\alpha t^2 \qquad (6\text{-}7)$$

(Continued on page 24)

ADDITIONAL INFORMATION

A particle moving in a circle of constant radius may experience a **tangential acceleration** $a_t = \alpha r$, if ω is changing with time. Even if ω is constant, a particle experiences the radial, or centripetal, acceleration, $a_c = \omega^2 r = \frac{v^2}{r}$, which is discussed in chapter 2.

The *cross product* $L = r \times p$ between two vectors r and p is a vector L which is perpendicular to both r and p. The direction of L is given by the right-hand rule (Fig. 6-2a). The magnitude of L is $L = rp \sin\theta$, where θ is the angle between r and p. If r and p are parallel, then $L = rp \sin 0 = 0$.

For a *rigid body* with moment of inertia I,

$$L = I\omega. \qquad (6\text{-}13)$$

Thus L is analogous to the linear momentum $p = mv$. If a body is continuous and not a set of discrete point masses, then Eq. 6-9a must be used to calculate the moment of inertia. If ρ is the density of the mass element dm, then the integral in Eq. 6-9a becomes,

$$I = \int_V r^2 \rho \, dV, \qquad (6\text{-}14)$$

where $\rho = m/v$, and r is the perpendicular distance from the volume element dV to the axis. If the axis is changed, then I also changes.

Example 1. Find the moment of inertia of a thin rectangular mass m with dimensions $a \times b \times c$, where c is the thickness. Consider the axis to be along the edge of length b. Then, $I = \int r^2 dm$ and $dm = \rho dV$, where the volume element dV is $bcdr$; hence,

$$I = \int_{r=0}^{a} r^2 \left(\frac{m}{abc}\right)(bcdr) = \frac{m}{3a} r^3 \Big|_0^a = \frac{ma^2}{3}.$$

The analogs between the angular quantities θ, ω, α, I, L, and τ, and the linear quantities s, v, a, m, p, and F (in that order) point out methods of solving problems in rotational mechanics.

Example 2. Consider a wheel of moment of inertia 0.1 kg m^2, which is rotating at 30 rad/sec. The wheel experiences a torque of 2 nt m, which tends to accelerate the rotation. How many revolutions does the

(Continued on page 24)

EXPLANATIONS

1. a $m = \dfrac{W}{g} = \dfrac{64 \text{ lb}}{32 \text{ ft/sec}^2} = 2$ slug. $I = mk^2 = (2 \text{ slug})$
   $(0.4 \text{ ft})^2 = 0.32$ slug ft^2.

   b $L = I\omega = (0.32 \text{ slug ft}^2)(5 \text{ rad/sec}) = 1.60$ slug ft^2/sec. Note that radians are dimensionless and are often omitted.

   c $\omega = \omega_0 + \alpha t, \alpha = \dfrac{\omega - \omega_0}{t} = \dfrac{5 \text{ rad/sec}}{10 \text{ sec}}$

   $= 0.5$ rad/sec^2.

2. a $\Delta z = \frac{1}{2} at^2$; since $v_0 = 0$. Thus $a = \dfrac{2\Delta z}{t^2} = \dfrac{2(250 \text{ cm})}{(1 \text{ sec})^2}$

   $= 500$ cm/sec^2

   $\alpha = \dfrac{a}{r} = \dfrac{500 \text{ cm/sec}^2}{10 \text{ cm}} = 50$ rad/sec^2.

   b First, find the torque on the wheel, $\tau = rF \sin 90° = rm_2 g$ (since $\mathbf{r} \perp \mathbf{F}$). $\tau = (10 \text{ cm})(50 \text{ gm})(980 \text{ cm/sec}^2) = 4.9 \times 10^5$ dy cm and since $\tau = I\alpha, I = \tau/\alpha =$
   $\dfrac{4.9 \times 10^5 \text{dy cm}}{50 \text{ rad/sec}^2} = 9.8 \times 10^3$ gm cm^2.

   c $k = \sqrt{\dfrac{I}{m_1}} = \sqrt{\dfrac{9.8 \times 10^3 \text{ gm cm}^2}{400 \text{ gm}}} = 4.96$ cm.

3. a $m = \dfrac{96 \text{ lb}}{32 \text{ ft/sec}^2} = 3$ slug; hence $I_b = mk_b^2 = (3 \text{ slug})$
   $(2 \text{ ft})^2 = 12$ slug ft^2, and $I_a = mk_a^2 = (3 \text{ slug})(1 \text{ ft})^2 =$
   3 slug ft^2. Thus, $L_b = I_b\omega_b = (12 \text{ slug ft}^2)(2\pi \text{ rad/sec}) = 24$ slug ft^2/sec, and $L_a = I_a\omega_a = (3 \text{ slug ft}^2)\omega_a$; but since $L_b = L_a, \omega_a = 8\pi$ rad/sec.

   b $K_{a \text{ rot}} = \frac{1}{2} I_a\omega_a^2 = \frac{1}{2} (3 \text{ slug ft}^2)(8 \pi \text{ rad/sec})^2$

   $= 96\pi^2$ ft lb, and

   $K_{b \text{ rot}} = \frac{1}{2} I_b\omega_b^2 = \frac{1}{2} (12 \text{ slug ft}^2)(2\pi \text{ rad/sec})^2$

   $= 24\pi^2$ ft lb; hence

   $\Delta K_{\text{rot}} = (96 - 24)\pi^2$ ft lb $= 72\pi^2$ ft lb. $= 713$ ft lb.

4. a $\omega_f^2 = \omega_0^2 + 2\alpha\theta$; hence, $0 = (3\pi \text{ rad/sec})^2 + 2\alpha(20)(2\pi \text{ rad/rev})$. Thus, $\alpha = -0.113\pi$ rad/sec^2.

   b $K_{b \text{ rot}} = \frac{1}{2} I_b\omega_b^2 = \frac{1}{2} (50 \text{ kg m}^2)$ $(3\pi \text{ rad/sec})^2 = 225\pi^2$ joule, and $K_{a \text{ rot}} = 0$. Thus $225\pi^2$ joule $= 2220$ joule are dissipated by the bearings in the form of heat.

Answers

0.32 slug ft^2	1 a
1.60 slug ft^2	b
0.5 rad/sec^2	c
50 rad/sec^2	2 a
9.8×10^3 gm cm^2	b
4.96 cm	c
8π rad/sec	3 a
$72\pi^2$ ft lb $= 713$ ft lb	b
-0.113π rad/sec^2	4 a
$225\pi^2$ joule $= 2220$ joule	b

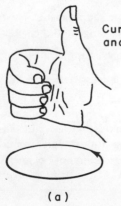

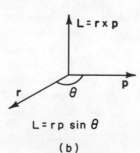

Curled fingers show that r is moving toward p and thumb points in direction of L.

$L = r \times p$

$L = r p \sin \theta$

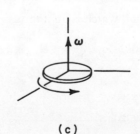

(a) (b) (c)

Fig. 6-2

$$\omega^2 = \omega_0^2 + 2\alpha\theta. \qquad (6\text{-}8)$$

The **moment of inertia** of a body is defined in terms of an axis of the body. If the body is composed of n masses m_i located at distance r_i from the axis, the moment of inertia I of the body is,

$$I = \sum_{i=1}^{n} m_i r_i^2. \qquad (6\text{-}9)$$

If the body is a continuous solid,

$$I = \int_m r^2 dm \qquad (6\text{-}9a)$$

The limits of the integral depend on the geometry of the body and the integration is performed over the entire body. The moment of inertia in rotational dynamics is analogous to the mass in ordinary dynamics. The rotational analog to Newton's second law is,

$$\tau_{\text{net}} = \sum_{i=1}^{n} \tau_i = I\alpha, \qquad (6\text{-}10)$$

where τ is the **torque**. The torque τ is defined as,

$$\tau = \mathbf{r} \times \mathbf{F}, \qquad (6\text{-}11)$$

where $\mathbf{r}$ is the *displacement vector*. Note that $\tau = \dfrac{d\mathbf{L}}{dt}$ is analogous to $\mathbf{F} = \dfrac{d\mathbf{p}}{dt}$.

The **radius of gyration** k of a solid of mass m and moment of inertia I is defined as,

$$k = \sqrt{\frac{I}{m}} \qquad (6\text{-}12)$$

The **rotational kinetic energy** $K_{\text{rot}} = \frac{1}{2}I\omega^2$, which is analogous to translational kinetic energy $K_{tr} = \frac{1}{2}mv^2$.

wheel make before it travels at 50 rad/sec and how long does it take to acquire this angular velocity?

First, solve for α: $\alpha = T/I = (2 \text{ nt m})/(0.1 \text{ kg m}^2) = 20 \text{ rad/sec}^2$. Using Eq. 6-8, where $\theta = \dfrac{\omega^2 - \omega_0^2}{2\alpha} = \dfrac{2500 - 900}{40} \text{ rad} = 40 \text{ rad}$. Thus, the wheel goes $40/2\pi = 20/\pi$ revolutions. The time can be found easily by using Eq. 6-6, where,

$$t = \frac{\omega - \omega_0}{\alpha} = \left(\frac{50 - 30}{20}\right) \text{ sec} = 1 \text{ sec}.$$

The moment of inertia of an object depends on the mass of the object and on the location of the axis. The following table catalogs the moment of inertia for some common shapes.

Shape (mass)	Location of Axis	Moment of Inertia
Hoop, radius R, and mass m	Axis of hoop (cylinder)	mR^2
	Tangent line to outer edge	$\frac{3}{2}mR^2$
Thin rod length L	Through center $\perp$ to length	$\frac{1}{12}mL^2$
Solid sphere	Through center	$\frac{2}{5}mR^2$
Spherical shell	Through center	$\frac{2}{3}mR^2$

If the axis of rotation for a solid is not through the center of mass, the moment of inertia can be found about this axis in terms of I_0, the moment of inertia for an axis through the center of mass; the two axes must be parallel; thus,

$$I = I_0 + mh^2, \qquad (6\text{-}15)$$

where m is the mass of the solid and h is the distance between the two axes. This relationship is called the *parallel-axis theorem*.

Example 3. Calculate the moment of inertia of a solid sphere about an axis tangent to the sphere. Since $I_0 = \dfrac{2mR^2}{5}$ and the two axes are parallel, the parallel-axis theorem can be used, where $h = R$ and $I = \frac{2}{5}mR^2 + mR^2 = \frac{7}{5}mR^2$.

The radius of gyration k is a measure of the size and shape of a solid. A hoop, for example, of mass m and radius k will have the same moment of inertia as the original solid.

7 STATICS AND MECHANICAL ADVANTAGE

1. In Fig. 7-1, calculate the tension, T_1 and T_2, in ropes 1 and 2, respectively. The stick weighs 20 lb.

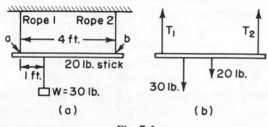

Fig. 7-1

2. Draw a free body diagram for the stick in Fig. 7-2a.
   a Find the tangential force exerted by the wall.
   b Find the tension on the rope.

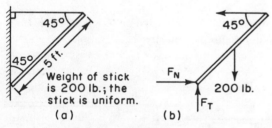

Fig. 7-2

3. A uniform 75 lb ladder shown in Fig. 7-3 is leaning against a frictionless wall. If the maximum frictional force the ground can provide is 100 lb, how far up the ladder can a 150 lb man climb before the ladder starts to slip?

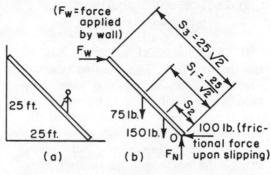

Fig. 7-3

4. A 200 lb box is pulled up a 20 ft inclined plane. The upper end of the plane is 12 ft higher than the bottom end. The force required to push the box up the plane is 160 lb.
   a Find the ideal mechanical advantage.
   b Find the actual mechanical advantage.
   c Find the mechanical efficiency.

5. In Fig. 7-4, a lever is used to lift a 30 lb weight 6 inches.
   a What distance must the 4 ft end move?
   b If the external force is 16 lb, what is the actual mechanical advantage?
   c What is the ideal mechanical advantage?
   d What is the mechanical efficiency?

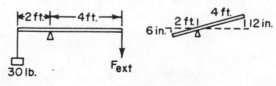

Fig. 7-4

1

2 a
  b

3

4 a
  b
  c

5 a
  b
  c
  d

25

BASIC FACTS

There are two conditions for equilibrium:

1) The sum of the forces, $\Sigma \mathbf{F}$ acting on a body is zero, and
2) The sum of the torques, $\Sigma \tau$ acting on a body is zero.

If the net force, $\Sigma \mathbf{F}$ acting on a body is zero, then from Newton's first law, the acceleration of that body is also zero, even though the body may be moving at a constant velocity. Furthermore, if the net torques $\Sigma \tau$ on a body is zero, the angular acceleration is also zero, even though the body may be rotating at a constant rate. This chapter will be concerned only with static bodies; that is, bodies with zero velocities and zero angular velocities.

All problems in statics require a *free body diagram*. The components of all the forces depend on the orientation of the axes in the diagram and the sign and magnitude of the torque produced by each force depend on the location of the origin. The rules for constructing a free body diagram are given in chapter 3.

An external force $\mathbf{F}_{ext}$ produces a displacement $\Delta \mathbf{x}_1$ on some part of a machine. The work delivered by the external force to the machine is $\mathbf{F}_{ext} \Delta \mathbf{x}_1$. The machine applies a force $\mathbf{F}_{int}$ which causes a displacement $\Delta \mathbf{x}_2$; thus, $\mathbf{F}_{int} \Delta \mathbf{x}_2$ is the work done by the machine in response to the external force. For an ideal, frictionless machine,

$$\mathbf{F}_{ext}\mathbf{x}_1 = \mathbf{F}_{int}\Delta \mathbf{x}_2; \qquad (7\text{-}1)$$

thus,

$$\frac{\mathbf{F}_{int}}{\mathbf{F}_{ext}} = \frac{\Delta \mathbf{x}_1}{\Delta \mathbf{x}_2} = R_I, \qquad (7\text{-}2)$$

where R_I is called the **ideal mechanical advantage**.

work = (Force)(displacement)

(Continued on page 28)

ADDITIONAL INFORMATION

A problem in statics can be solved by using any choice of coordinate axes (x, y, and z directions) and of an origin. Solutions are simplest, however, if the origin is chosen so that as many forces as possible are directed through it.

Torques were defined in chapter 6 as $\tau = \mathbf{r} \times \mathbf{F}$ and $\tau = rF \sin \theta$; thus τ is perpendicular to both $\mathbf{r}$ and $\mathbf{F}$ according to the right-hand rule. In this chapter, the forces on a body and the distances from the axes will be drawn in the plane of the paper. The torque will therefore be directed into or out of the paper. Torques which tend to turn a body *counterclockwise* about the chosen origin are directed out of the paper, according to the right-hand rule. Such a torque is *positive* for the purposes of calculation. A torque which tends to turn a body *clockwise* about the chosen origin is directed into the paper. Such a torque is *negative*.

In order to solve a statics problem, we must substitute the known forces and distances into the equations $\Sigma \mathbf{F} = 0$ and $\Sigma \tau = 0$, the conditions for equilibrium, and solve these equations for the unknown forces and distances.

Example 1.   The pole in Fig. 7-5 is in static equilibrium because its velocity, acceleration, angular ve-

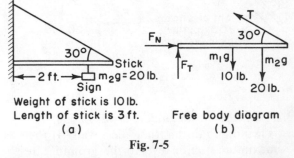

Weight of stick is 10 lb.
Length of stick is 3 ft.
(a)

Free body diagram
(b)

Fig. 7-5

locity, and angular acceleration are all zero. This is shown in the free body diagram. Only the pole is drawn; the wall, sign, and wire are replaced by the forces they exert on the pole. The origin is chosen at the point at which the pole joins with the wall. The axes are chosen both along the pole and perpendicular to it. The forces exerted on the pole by the wall are resolved into two components F_N, which is *normal* to the wall, and F_T, which is *tangential* to the wall; the latter keeps the pole from slipping. The weight of

(Continued on page 28)

EXPLANATIONS

1. $\Sigma F_x = 0$; since there are no forces in the x direction, $\Sigma F_y = T_1 + T_2 - 20 \text{ lb} - 30 \text{ lb} = 0$, thus $T_1 + T_2 = 50$ lb. The origin for calculating the torques may be chosen anywhere; choosing the origin at point a, then $\Sigma \tau = -m_1 g \Delta x_1 - m_2 g \Delta x_2 + T_2 \Delta x_3 = -(30 \text{ lb})(1 \text{ ft}) - (20 \text{ lb})(2 \text{ ft}) + T_2(4 \text{ ft}) = 0$, thus $T_2 = 17.5$ lb and $T_1 = 30 \text{ lb} - T_2 = 12.5$ lb.

2. a $\Sigma F_x = F_N - T = 0$; $\Sigma F_y = F_T - mg = 0$, thus $F_T = mg = 200$ lb.

   b Choosing the origin at the point where the stick meets the wall, $\Sigma \tau = -mg \Delta x \sin \theta + T \Delta_x \sin \theta = -(200 \text{ lb})(2.5 \text{ ft}) \sin 45° + T(5 \text{ ft}) \sin 45° = 0$, thus $T = 100$ lb.

3. Choosing the origin at point 0 in Fig. 7-3b, $\Sigma F_x = -F_f + F_w = 0$, thus $F_w = F_f = 100$ lb just before the ladder slips.
   $\Sigma \tau = F_1 s_1 \sin 45° + F_2 s_2 \sin 45° - F_3 s_3 \sin 45°$
   $= (75 \text{ lb})(25 \text{ ft})(0.707) + (150 \text{ lb}) s_2 (0.707) -$
   $(100 \text{ lb})(50 \text{ ft})(0.707) = 0,$
   thus $s_2 = 14.7$ ft.

4. a $F_{ext} = 160$ lb, $\Delta x_1 = 20$ ft; $F_{int} = 200$ lb, $\Delta x_2 = 12$ ft. Therefore,

   $$R_I = \frac{\Delta x_1}{\Delta x_2} = \frac{20 \text{ ft}}{12 \text{ ft}} = 1.67.$$

   b $$R_A = \frac{F_{int}}{F_{ext}} = \frac{200 \text{ lb}}{160 \text{ lb}} = 1.25.$$

   c $$e = \frac{R_A}{R_I} = \frac{1.25}{1.67} = 0.75.$$

5. a From the geometry in Fig. 7-4, the 4 ft end must rise 12 in. when the 2 ft end is lowered 6 in.

   b $$R_A = \frac{30 \text{ lb}}{16 \text{ lb}} = 1.88.$$

   c $$R_I = \frac{12 \text{ in}}{6 \text{ in}} = 2.0.$$

   d $$e = \frac{R_A}{R_I} = \frac{1.88}{2.0} = 0.94.$$

Answers

$T_1 = 12.5$ lb, $T_2 = 17.5$ lb	1
200 lb	2 a
100 lb	b
14.7 ft	3
1.67	4 a
1.25	b
0.75	c
12 in.	5 a
1.88	b
2.0	c
0.94	d

The **actual mechanical advantage** R_A is given by,

$$R_A = \frac{F_{int}}{F_{ext}}. \qquad (7\text{-}2a)$$

For a frictionless machine, the **mechanical efficiency** e is,

$$e = \frac{R_A}{R_I} = 1 \text{ (unity).} \qquad (7\text{-}3)$$

However, for an actual machine, $e < 1$.

the sign m_2g has been drawn down from the *center of mass* of the sign and the weight of the pole m_1g has been drawn down from the center of mass of the pole, which is the midpoint, in this case, 1.5 ft from the wall. The tension **T** of the wire is drawn along the line of the wire. The tension **T** and the forces $\mathbf{F}_N$ and $\mathbf{F}_T$ are found by applying the conditions of equilibrium to this problem. First,

$$\Sigma F_x = F_N - T\cos 30° = 0,$$
$$\Sigma F_y = F_t - m_1g - m_2g + T\sin 30°;$$

hence, $F_N = T\cos 30°.$

There are three unknowns T, $\mathbf{F}_N$, and $\mathbf{F}_T$. Applying the second condition of equilibrium,

$$\Sigma\tau = -m_1g\,\Delta x_1 - m_2g\,\Delta x_2 + (T\sin 30°)\,x_3$$
$$= -(-10\text{ lb})(1.5\text{ ft}) + (-20\text{ lb})(2\text{ ft})$$
$$+ \left(T\,\tfrac{1}{2}\right)(3\text{ ft}) = 0$$

Solving for T, $T = 16.7$ lb; hence $F_N = 14.5$ lb and $F_T = 38.4$ lb.

The basic equation for mechanical efficiency e is

$$e = \frac{\text{work performed by machine}}{\text{work put into machine}}$$

For a frictionless machine $e = 1$; this means that the machine is 100% efficient.

The ideal mechanical advantage R_I can be measured from the geometry of the machine. The actual mechanical advantage R_A, however, requires knowledge of the amount of friction present in bearings or other moving parts. If $e = 1$ and $R_A = R_I$, all of the work performed by the external force is transformed into useful work. However, if $e < 1$, some of the work provided by the external force is dissipated in the form of heat.

8

GRAVITATION

1. A 160 lb man is standing on a set of scales in an elevator. When the elevator starts to move downward, he reads 128 lb on the scale.
   a What is the acceleration of the elevator?
   b If the elevator moves upward with the same acceleration, what will the man read on the scale?

2. The mass of the moon is 0.012 times the mass of the earth M_e and its radius is 0.27 times the earth's radius R_e.
   a What is the acceleration of gravity on the moon (in m/sec^2)?
   b How much would a 50 kg rock weigh on the moon?

3. A 2 kg mass is attached to a horizontal light spring. When the mass is pulled with a force of 4.0 nt, it moves 2 cm.
   a What is the spring constant?
   b If the mass is released, describe its subsequent motion and what is the period.

4. Calculate the gravitational force between two 100 kg masses 1 in apart.

5. What is the period of a 9.8 cm pendulum on earth?

6. Find the period of a satellite moving near the earth's surface. (The earth's radius $R_e = 6.38 \times 10^6$ m.)

7. Find the period of a satellite orbiting at a height of $2R_e$ above the earth's surface.

8. Calculate the mass of the earth. (The earth's radius $R_e = 6.38 \times 10^6$ m.)

9. Calculate the potential energy of a 100 kg satellite 6.2×10^5 kg above the earth's surface. (The earth's mass $M_e = 6 \times 10^{24}$ kg and the earth's radius R_e is given in problem 8.)

1	a
	b
2	a
	b
3	a
	b
4	
5	
6	
7	
8	
9	

BASIC FACTS

Newton's law of gravitation states that,

$$F_{12} = \frac{-GM_1M_2}{R^2}, \qquad (8\text{-}1)$$

where F_{12} is the gravitational force directed from mass M_1 to mass M_2, R is the distance between the centers of mass of the two objects (masses), and G is **Newton's constant**, which is equal to 6.67×10^{-8} dy cm^2/gm^2 = 6.67×10^{-11} nt m^2/kg^2.

The **gravitational potential energy** of a mass m above the earth's surface is,

$$U = mgz, \qquad (8\text{-}2)$$

where z is the height of mass m above the earth's surface. When larger distances are dealt with, a more general formula must be used,

$$U = \frac{-GM_em}{R}, \qquad (8\text{-}3)$$

where R is the distance from the center of the earth to m and M_e is the mass of the earth. When Eq. 8-2 is used, U is chosen to be zero at $z = 0$ (at the surface of the earth). When Eq. 8-3 is used, $U = 0$ at $R = \infty$. (∞ stands for "infinity.")

Kepler's Laws.
1. The orbit of every planet around the sun forms an ellipse with the sun at one of the foci.
2. The radius vector between the sun and a planet sweeps out equal areas in equal times.
3. For all the planets, the squares of their periods of revolution T are proportional to the cubes of their mean, or average, distances R from the sun. For any two planets, then,

$$\frac{T_1^2}{R_1^3} = \frac{T_2^2}{R_2^3}. \qquad (8\text{-}4)$$

(Continued on page 32)

ADDITIONAL INFORMATION

The gravitational force on a body of mass m at an arbitrary value of R, the distance from the earth, can often be found by using the gravitational force on the body on the earth's surface; that is,

$$mg = \frac{GM_em}{R_e^2}, \text{ hence, g} = \frac{GM_e}{R_e^2},$$

which is derived from Newton's law of gravitation. Eq. 8-1 can be used to calculate the gravitational force on the body, if R is expressed in terms of R_e, the radius of the earth.

Example 1.   Find the gravitational force F_{grav} on a 1 slug mass a distance of 1 R_e from the surface of the earth. Since R_e is the distance from the mass to the center of the earth, $R = 2R_e$, and,

$$F_{grav} = \frac{GM_em}{R^2} = \frac{GM_em}{(2R_e)^2} = \frac{1}{4}\,m\left(\frac{GM_e}{R_e^2}\right)$$

$$= \frac{1}{4}\,mg = \frac{1}{4}\,(1\text{ slug})(32\text{ ft sec}^2) = 8\text{ lb.}$$

Kepler's first law can be derived by using Newton's laws. The second law is equivalent to the law of conservation of angular momentum; that is, when a planet is close to the sun, it moves faster than it does when it is far away from the sun. The angular momentum, $\mathbf{L} = \mathbf{r} \times \mathbf{p}$, is constant and proportional to the amount of area swept out per unit time. Kepler's laws apply to any system with small bodies revolving around large ones. The moons of Jupiter, for example, and the artificial satellites of the earth obey Kepler's laws. Kepler's third law relates the periods and average radii of two planets revolving around the sun, but it does not relate the periods or average radii of a planet and a satellite, since this law only relates bodies having their orbits around the same object.

Example 2.   What is the period T_N of Neptune's revolution, if its average distance to the sun is 30 times the distance R_e' from the earth to the sun? Remembering that the earth's period T_e is 1 year; thus,

$$\frac{T_N^2}{(30\,R_e')^2} = \frac{T_E^2}{R_e'^2}; \quad \frac{T_N^2}{27,000\,R_e'^2} = \frac{1\text{ yr}}{R_e'^2};$$

hence, $T_N = 27,000\text{ yr} = 164.5\text{ yr.}$

(Continued on page 32)

EXPLANATIONS

1. a The acceleration is downward; thus, $W' = m(g - a)$.

   $m = \dfrac{W}{g} = \dfrac{160\ \text{lb}}{32\ \text{ft/sec}^2} = 5$ slug; $W' = 128$ lb $=$
   $(5\ \text{slug})(32\ \text{ft/sec}^2 - a)$, thus $a = 6.4\ \text{ft/sec}^2$.

   b $W' = m(g + a) = (5\ \text{slug})(32 + 6.4)\ \text{ft/sec}^2 = 192$ lb.

2. a $g' = \dfrac{GM_m}{R_m^2}$, where M_m is the mass of the moon, and

   R_m is its radius; hence $g' = \dfrac{G(0.012 M_e)}{(0.27\ R_e)^2} = \dfrac{0.012}{0.0729}\ \dfrac{GM_e}{R_e^2}$

   $= 0.164\ g = 1.61\ \text{m/sec}^2.\ (g = 9.8\ \text{m/sec}^2.)$

   b $W = mg' = (5\ \text{kg})(1.61\ \text{m/sec}^2) = 8.05$ nt.

3. a $k = -\dfrac{F}{x} = \dfrac{-(-4\ \text{nt})}{(0.02\ \text{m})} = 200\ \text{nt/m}.$ (k is always positive.)

   b The mass oscillates about its equilibrium point with a period T, where,

   $$T = 2\pi\sqrt{m/k} = 2\pi\sqrt{\dfrac{2\ \text{kg}}{200\ \text{nt/m}}} = 0.2\pi\ \text{sec}.$$

4. $F = \dfrac{-GM_1 M_2}{R^2} = \dfrac{-(6.67 \times 10^{-11}\text{nt m}^2/\text{kg}^2)(100\ \text{kg})^2}{(1\ \text{m})^2}$

   $= -6.67 \times 10^{-7}$ nt.

5. $T = 2\pi\sqrt{\dfrac{L}{g}} = 2\pi\sqrt{\dfrac{9.8\ \text{cm}}{980\ \text{cm/sec}^2}} = 0.2$ sec.

6. The sum of the forces on a low flying satellite is mg; the

   acceleration is $a_c = \dfrac{v^2}{R_e} = \dfrac{4\pi^2 R_e}{T^2}.\ mg = ma_c$; therefore,

   $T^2 = \dfrac{4\pi R_e}{g} = \dfrac{4\pi^2(6.38 \times 10^6 \text{m})}{9.8\ \text{m/sec}^2} = 25.7 \times 10^6\ \text{sec}^2.$ $T = 5.07 \times 10^3$ sec

7. $\dfrac{T_1^2}{R_1^3} = \dfrac{T_2^2}{R_2^3}$; then $\dfrac{(5 \times 10^3 \text{sec})^2}{R_e^3} + \dfrac{T_2^2}{(3\ R_e)^3}$, thus

   $T_2^2 = (27)(25 \times 10^6)\ \text{sec}^2$, and $T_2 = 26 \times 10^3$ sec

8. $g = \dfrac{GM_e}{R_e^2}$; thus, $M_e = \dfrac{gR_e^2}{G} = \dfrac{(9.8\ \text{m/sec}^2)(6.38 \times 10^6 \text{m})^2}{(6.67 \times 10^{-11}\ \text{nt m}^2/\text{kg}^2)}$

   $= 59.6 \times 10^{23}$ kg.

9. $R = (6.38 + 0.62) \times 10^6\ \text{m} = 7.00 \times 10^6\ \text{m}$; thus,

   $U = \dfrac{-GM_e m}{R}$

   $= \dfrac{-(6.67 \times 10^{-11}\ \text{nt m}^2/\text{kg}^2)(6 \times 10^{24})(10^2)\text{kg}^2}{(7 \times 10^6\ \text{m})}$

   $= -5.72 \times 10^9$ joule.

Answers

$6.4\ \text{ft/sec}^2$	1 a
192 lb	b
$1.61\ \text{m/sec}^2$	2 a
8.05 nt	b
200 nt/m	3 a
The mass oscillates; 0.2π sec	b
-6.67×10^{-7} nt	4
0.2 sec	5
5.07×10^3 sec	6
26×10^3 sec	7
59.6×10^{23} kg	8
-5.72×10^9 joule	9

Simple Harmonic Motion (SHM). If the acceleration a of an object is proportional to its position x, then,

$$a = -\sigma^2 x, \qquad (8\text{-}5)$$

where σ^2 is a positive constant and the object moves back and forth. The motion is **periodic** because it repeats itself once during every regular time interval, which is called a *period*. The **frequency** f is the number of complete vibrations per unit time; thus,

$$f = \frac{1}{T}, \qquad (8\text{-}6)$$

hence $\quad T = \dfrac{2\pi}{\sigma}, \qquad (8\text{-}6a)$

and $\quad x = A \sin(2\pi f t + \phi), \qquad (8\text{-}7)$

where ϕ is the **phase angle** and A is the **amplitude**, or the maximum value of x. If the object starts at $x = 0$ when $t = 0$, then $\phi = 0$. Two common examples of SHM are the oscillations of a spring and the swinging of a pendulum. The quantity σ (Eq. 8-5) is $\sigma^2 = k/m$ for the spring, where k is the spring constant and m is the mass connected to the spring. $\sigma^2 = g/L$ for a pendulum, where L is the length of the pendulum and g is the acceleration of gravity.

The apparent weight W' of an object in an accelerating (noninertial) reference frame is,

$$W' = m \,|\, g\mathbf{j} - \mathbf{a}\,|, \qquad (8\text{-}8)$$

where $\mathbf{a}$ is the acceleration of the frame.

A reference frame may be stationary, moving at a constant velocity, or accelerating. Measurements of position, velocity, and acceleration are taken relative to a reference frame. If the measurements of acceleration of a body agree with Newton's first and second laws; that is, if $\Sigma \mathbf{F} = m\mathbf{a}$, then the reference frame is an inertial one. Inertial frames are either at rest or moving at constant velocities; Newton's first two laws do not hold in accelerating reference frames. In Eq. 8-5, if the acceleration is upward, the magnitude of the apparent weight is $W' = m(g + a)$; thus, $W' > mg$. In an elevator which is starting upward or in one that is stopping after having moved down, the acceleration is upward; if the reference frame is accelerating downward, the magnitude of the apparent weight is $W' = m(g - a)$ and is obviously less than mg. In an elevator which is starting down or in one which is stopping after having moved upward, the acceleration is downward.

Whenever the acceleration of an object has the opposite sign from its position, the particle will oscillate. Thus, if the position is negative, the particle accelerates toward the positive direction. When the particle passes through zero in moving from a positive to a negative point, the sign of the acceleration changes from negative to positive. SHM is a special case of oscillatory motion, and the position of an object undergoing SHM is given by the sine function, $x = A \sin(\sigma t + \phi)$. Notice that at $t = 0$, $x = A \sin \phi$; the motion repeats itself once during every period T. A mass m oscillating at the end of a spring is accelerating, since $F = ma = -kx$; thus, $a = -kx/m$ and $k/m = \sigma^2$ (Fig. 8-1). The equation for x as a function of time is,

$$x = A \sin\left[\sqrt{\frac{k}{m}}\, t + \phi\right],$$

and the period is $T = \dfrac{2\pi}{\sigma} = 2\pi\sqrt{\dfrac{m}{k}}$. If the amplitude of a pendulum is small, the distance x from the equilibrium position (the place it would hang if it were not moving) of the pendulum bob to its position at any time is along a curved path, thus SHM is required to move a distance x in time. Since $F = ma = -mgx/L$, $a = -gx/L$, so that $\sigma = \sqrt{\dfrac{g}{L}}$ (Fig. 8-2).

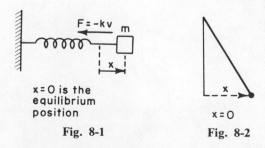

x=0 is the equilibrium position

Fig. 8-1

x = 0

Fig. 8-2

9

1. A piece of metal weighs 50 nt. When immersed in water, it has an apparent weight of 30 nt. (The density of water $\rho_w = 1 \text{ gm/cm}^3 = 1000 \text{ kg/m}^3$.)
   a What is the volume of the metal?
   b What is the density of the metal?
   c What is the specific gravity of the metal?

2. Piston A in Fig. 9-1 has a diameter of 6 inches; piston B has a diameter of 12 inches.
   a In order to lift a 500 lb weight on piston B, how much force must be applied to piston A?
   b How far down must piston A be pushed in order to lift the weight 5 in.?

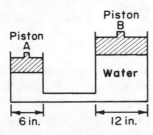

Fig. 9-1

3. The 5000 gm piston shown in Fig. 9-2 has a surface area of 10 cm². The liquid inside has a density of 0.8 gm/cm³. Find the pressure 5 cm below the piston. (Atmospheric pressure $P_0 = 10^6 \text{ dy/cm}^2$.)

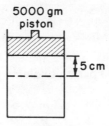

Fig. 9-2

4. How deep into the water must a diver go before the total pressure on him is twice the atmospheric pressure? ($P_0 = 10^6 \text{ dy/cm}^2$.)

5. The tube shown in Fig. 9-3 contains water, the velocity at point 1 is 1000 cm/sec, the radius at point 2 is 30 cm, and the radius at point 1 is 1 cm. (The density of water $\rho_w = 1 \text{ gm/cm}^3$ and the atmospheric pressure $P_0 = 10^6 \text{ dy/cm}^2$.)
   a Find the velocity at point 2.
   b Find the pressure at point 1, if the pressure at point 2 is atmospheric.

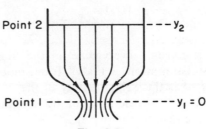

Fig. 9-3

1 a
   b
   c

2 a
   b

3

4

5 a
   b

33

BASIC FACTS

A **fluid** is any substance that changes its shape to conform to its container; thus gases and liquids are fluids. *Hydrostatics* is the study of fluids at rest, and *hydrodynamics* is the study of fluids in motion.

The **density** ρ of a material is defined as its mass m per unit volume V; thus,

$$\rho = \frac{m}{V}. \qquad (9\text{-}1)$$

The **specific gravity** S of a substance is the ratio of the density ρ of the substance to the **density of water** $\rho_w = 1 \text{ gm/cm}^3$,

$$S = \frac{\rho}{\rho_w}, \qquad (9\text{-}2)$$

where S is dimensionless.

Pressure P is defined as force per unit area. If a force F exerted by the fluid is constant over an area A, then $P = F/A$. The hydrostatic pressure of a fluid increases with depth, so that the difference between any two pressure points,

$$P_2 - P_1 = -\rho g(y_2 - y_1), \qquad (9\text{-}3)$$

if ρ is constant everywhere in the fluid. This assumption is valid for liquids. Here ρg is the **weight density** of the fluid (g being the acceleration of gravity) and y_2 is the elevation above y_1.

Consider an object immersed in a fluid of density ρ. It displaces a volume V of fluid. **Archimedes' principle** states that the fluid exerts an upward force called the **buoyancy force** F_b; thus,

$$F_b = \rho g V, \qquad (9\text{-}4)$$

The path of a small segment of fluid is called a **line of flow**. When the path is constant with time, the flow is called

(Continued on page 36)

ADDITIONAL INFORMATION

At the surface of a liquid $y_2 = 0$ and $P_2 = P_0$, the atmospheric pressure, but at a point $y_1 = h$ below the surface,

$$P = P_0 + \rho g h. \qquad (9\text{-}7)$$

This result follows from Eq. 9-3 where the pressure may be exerted on a diver, a balloon, or the walls of the container at h. The pressure is independent of the shape of the object or wall.

The **buoyancy force** $F_b = \rho g V$ is equal to the weight of the liquid displaced; for example, a rock with a volume of 10 cm^3 put into water experiences an upward force $F_b = \rho_w g h = (1 \text{ gm/cm}^3)(980 \text{ cm/sec}^2)(10 \text{ cm}^3) = 9.8 \times 10^3$ dy (Fig. 9-5). Notice that the

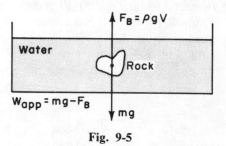

Fig. 9-5

density of the rock is not important; only the volume of water displaced matters. If the rock is only half submerged, then $F_b = 4.9 \times 10^3$ dy, or half of the original force. If an object floats in a liquid, only part of its volume is submerged.

Example 1.   A wooden cube 5 cm on every side has a density of 0.8 gm/cm^3. How deep will it be submerged in water? Here, Archimedes' principle is used to find the depth. Since the block is at equilibrium, it does not rise or sink and the buoyancy force must equal the weight of the block. Thus, $\rho_w g V_w = \rho_b g V_b$, where ρ_b and ρ_w are the densities of the block and of the water, respectively, and V_w is the volume of the water displaced and V_b is the total volume of the block. Since $\rho_b = 0.8\rho_w$ and $V_w = 0.8\ V_b$, the block is 80% submerged.

The buoyancy force is sometimes called the **apparent loss of weight**. If an object sinks in a liquid, the apparent weight W_{app} of the object is less than mg; hence,

$$W_{app} = mg - F_b. \qquad (9\text{-}8)$$

(Continued on page 36)

EXPLANATIONS

1. a $W_{app} = mg - F_b = mg - \rho_w g V$; thus, 30 nt = 50 nt $-$ (1000 kg/m^3)(9.8 m/sec^2) V and $V = 2.04 \times 10^{-3}$ m^3.

   b $\rho = \dfrac{m}{V}$ and $m = \dfrac{mg}{g} = \dfrac{50 \text{ kg}}{9.8 \text{ m/sec}^2} = 5.1$ kg; hence,

   $\rho = \dfrac{5.1 \text{ kg}}{2.04 \times 10^{-3} \text{ m}^3} = 2.5 \times 10^3$ kg/m^3.

   c $S = \dfrac{\rho}{\rho_w} = \dfrac{2.5 \times 10^3 \text{ kg/m}^3}{10^3 \text{ kg/m}^3} = 2.5$.

2. a From Pascal's law, $P_A = P_B$; thus, $P_B = \dfrac{F_B}{A_B} =$ $\dfrac{F_B}{\pi(6 \text{ in.})^2} = \dfrac{500 \text{ lb}}{\pi(6 \text{ in.})^2}$ and $P_A = \dfrac{F_A}{R_A^2} = \dfrac{F_B}{(3 \text{ in.})^2}$. Equating P_A to P_B, $\dfrac{500 \text{ lb}}{36\pi \text{ in.}^2} = \dfrac{F_A}{9\pi \text{ in.}^2}$, thus $F_A = 125$ lb.

   b The velocities of the pistons A and B are, respectively, $v_A = \dfrac{\Delta y_A}{\Delta t}$ and $v_B = \dfrac{\Delta y_B}{\Delta t}$, where y_A is the increase in height of piston A in time Δt; from the equation of continuity, $\dfrac{A_A \Delta y_A}{\Delta t} = \dfrac{A_B \Delta y_B}{\Delta t}$. Therefore, 9π in.^{2} $y_A = 36\pi$ in.^{2} (5 in.) and $y_A = 20$ in.

3. The piston exerts a force mg on the liquid and a corresponding pressure $P = F/A = mg/A$. The total pressure of the liquid is thus $P = P_0 + \rho g h + \dfrac{mg}{A} = 10^6$ dy/cm^2 $+$ (0.8 gm/cm^3)(980 cm/sec^2)(5 cm) $+$ $\dfrac{(5000 \text{ gm})(980 \text{ cm/sec}^2)}{10 \text{ cm}^2} = 1.49 \times 10^6$ dy/cm^2.

4. $P = P_0 + \rho g h = 2 P_0$, where $P_0 = 10^6$ dy/cm^2, the atmospheric pressure; thus $\rho g h = P_0$, and $h = \dfrac{P_0}{\rho g} =$ $\dfrac{10^6 \text{ dy/cm}^2}{(1.0 \text{ gm/cm}^3)(980 \text{ cm/sec}^2)} = 1020$ cm.

5. a $A_2 v_2 = A_1 v_1$, thus $\pi(30 \text{ cm})^2 v_2 = \pi(1 \text{ cm})^2$ (10^3 cm/sec) and $v_2 = 1.1$ cm/sec.

   b From Bernoulli's equation, $\frac{1}{2}\rho(v_1^2 - v_2^2) = (P_2 - P_1) +$ $\rho g(y_2 - y_1)$; thus, since $y_2 - y_1 = 10^3$ cm, $\frac{1}{2}(1 \text{ gm/cm}^3)(10^6 - 1.1)cm^2/sec^2 = P_0 - P_1 +$ $(1 \text{ gm/cm}^3)(980 \text{ cm/sec}^2)(10^3 \text{ cm})$. Taking $(10^6 - 1.1)$ cm^2/sec$^2 \cong 10^6$ cm^2/sec^2 and solving for P_1 using the fact that $P_0 = 10^6$ dy/cm^2, the atmospheric pressure, $P_1 = (10^6 + 0.98 \times 10^6 - 0.5 \times 10^6)$ dy/cm$^2 = 1.49 \times 10^6$ dy/cm^2.

Answers

2.04×10^{-3} m^3	1 a
2.5×10^3 kg/m^3	b
2.5	c
125 lb	2 a
20 in.	b
1.49×10^6 dy/cm^2	3
1020 cm	4
1.1 cm/sec	5 a
1.49×10^6 dy/cm^2	b

steady or *stationary flow*. A **streamline** is a curve on which the tangent at every point represents the direction of the fluid velocity. A **tube of flow** is a tube formed by streamlines. The cross-sectional area of a tube of flow may vary along the tube (Fig. 9-4).

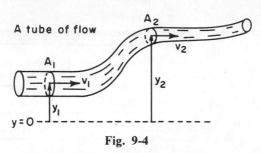

A tube of flow

Fig. 9-4

The **equation of continuity** relates the velocity of an incompressible fluid at different points in a tube of flow. If v_1 and A_1 are, respectively, the *velocity* of the fluid and the cross-sectional *area* of the tube at a point along a tube of flow, v_2 and A_2 are the velocity and area at a different point along the same tube. The **equation of continuity** is,

$$A_1 v_1 = A_2 v_2; \qquad (9\text{-}5)$$

thus Av is constant for any tube of flow.

Bernoulli's equation relates the pressure P, velocity v, and height y (or h) of a fluid at different points in the tube of flow; thus,

$$P_1 + \rho g y_1 + \tfrac{1}{2} \rho v_1^2 = P_2 + \rho g y_2 + \tfrac{1}{2} \rho v_2^2$$
$$(9\text{-}6)$$

where P is the absolute pressure; thus, $P + \rho g h + \tfrac{1}{2} \rho v^2$ is a constant for a given tube of flow.

The units of pressure are dy/cm^2, nt/m^2, and lb/in^2. Sometimes pressure is given in atmospheres; for example, 2 atmospheres of pressure means twice the normal atmospheric pressure. Pressure is also given in cm of mercury, however cm is not a unit of pressure: 1 atmosphere is about 76 cm of mercury.

Pascal's law states that any pressure applied to an enclosed liquid is transmitted throughout the entire liquid and to the walls of the containing vessel. This principle is necessary for calculating the forces exerted by a hydraulic press.

The equation of continuity is really an expression of the conservation of mass; it is often used in conjunction with Bernoulli's equation, which is equivalent to the law of conservation of energy, especially when the flow is in a tube.

An airplane wing experiences a lifting force l because the air pressure above the wing is smaller than the pressure below the wing. This pressure difference can be calculated using Bernoulli's equation (Eq. 9-6). The pressures and velocities above and below the wing are related by,

$$\tfrac{1}{2} \rho (v_1^2 - v_2^2) = (P_2 - P_1) + \rho g (y_2 - y_1).$$
$$\text{(See Eq. 9-3.)}$$

The term $(y_2 - y_1)$ is negligible for normal wings, and $F_l = A(P_2 - P_1)$, where A is the area of the wing; also $F_l = \dfrac{A}{2} \rho (v_1^2 - v_2^2)$.

Example 2.   Find the velocity of the water escaping from a hole at the bottom of a dam. The dam is 10 m high and the atmospheric pressure $P_0 = 10^6$ dy/cm². $P_1 = P_2$ because both pressures are open to the atmosphere; $y_1 = 0$, $y_2 = -10^3$ cm; and $v_1 = 0$, since the level moves very slowly; thus,

$$\tfrac{1}{2} \rho v_2^2 - \rho g y_2 = 0; \text{ hence, } v_2 = \sqrt{2 g y_2}$$
$$= \sqrt{2(980 \text{ cm/sec}^2)(10^3 \text{ cm})} =$$
$$= 1.4 \times 10^3 \text{ cm/sec}.$$

10

MECHANICAL AND THERMAL PROPERTIES OF MATTER

SELF-TEST

1. A steel rod 4 m long and 2 cm² in cross section is subjected to a tension of 3×10^9 dy. Young's modulus for the steel is 2.0×10^{12} dy/cm². How much will the rod stretch?

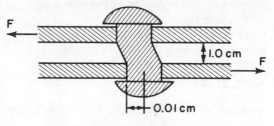

Fig. 10-1

2. If 5 m³ of carbon disulfide ($k = 64 \times 10^{-11}$ m²/nt) is subjected to a pressure of 1000 nt/m², what will be the change in its volume?

3. A brass rivet with a cross-sectional area of 5 cm² is mounted as shown in Fig. 10-1. This arrangement is used to measure the tangential force. If the shear modulus is known to be 0.36×10^{12} dy/cm and $x = 0.01$ cm, what is the tangential force?

4. Three moles of an ideal gas are in a piston. The initial pressure is atmospheric (10^5 nt/m²) and the initial volume is 2 m³.
   a   Calculate the temperature of the gas.
   b   If the gas is compressed isothermally (at a constant temperature) to 1 m³, what is the new pressure?
   c   If the temperature is doubled at constant pressure, what happens to the volume?

BASIC FACTS

When forces are applied to a solid body, the body tends to distort slightly. A force per unit area is called a **stress** and the relative change in a dimension or shape of a body subject to a stress is called a **strain**. A force applied to a surface has a normal component F_N (perpendicular to the surface) and a tangential component F_T (along the surface). Each type of stress causes a strain which is generally proportional to the stress.

The **tension stress** is the normal force per unit surface area F_N/A tending to elongate a body (Fig. 10-2). The cor-

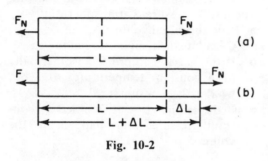

Fig. 10-2

responding strain is the ratio $\Delta L/L$, where ΔL is the increase in length L. The coefficient of proportionality is **Young's modulus** Y, where,

$$F_N/A = \frac{Y\Delta L}{L}. \qquad (10\text{-}1)$$

The **compression stress** is the normal force per unit area F_N/A tending to compress a body (Fig. 10-3). The decrease

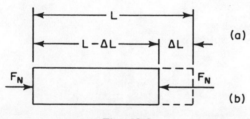

Fig. 10-3

(Continued on page 40)

ADDITIONAL INFORMATION

The equation of state for an ideal gas (Eq. 10-5) describes the behavior of real gases at low pressures and high temperatures; each real substance, however, has its own equation of state which describes its behavior for all temperatures and pressures. The molecules of an ideal gas do not interact; therefore, liquefaction and solidification are impossible for an ideal gas.

A real substance generally has three phases, the **gas phase** (or *vapor phase*), the **liquid phase,** and the **solid phase.** In Fig. 10-5, P versus T is shown for a real

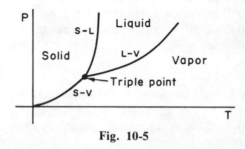

Fig. 10-5

substance. The lines S-L, L-V, and S-V are the solid-liquid, liquid-vapor, and solid-vapor lines, respectively. The S-V line is a graph of the **sublimation temperature** versus the pressure. If the pressure-temperature point (T, P) falls on this curve, the substance can exist as a solid or a gas at equilibrium. In this case, an amount of heat called the **latent heat of sublimation** must be added to change 1 gram of the substance from the gas phase to the solid phase. This change occurs with no change of temperature.

The L-V line is a plot of the *boiling point* versus the pressure. If the point (T, P) lies on this curve, the substance can exist in the liquid or in the vapor phase. In this state, an amount of heat called the **latent heat of vaporization** must be added to 1 gram of liquid to change it into 1 gram of gas at the same temperature.

The S-L line is a plot of the **melting temperature** which depends on pressure. If the slope is positive, the substance freezes when pressure is applied to the liquid phase. Although the S-L line for most substances is positive, a few substances, like water, have negative S-L slopes and the solid phase melts if sufficient pressure is applied.

(Continued on page 40)

EXPLANATIONS

1. $\Delta L = \dfrac{TL}{Y} = \dfrac{(3 \times 10^9 \text{ dy})(400 \text{ cm})}{2 \times 10^{12} \text{ dy/cm}} = 0.6 \text{ cm}.$

2. $\dfrac{\Delta V}{V} = -kP$; thus,

   $\Delta V = -kPV = -(64 \times 10^{-11} \text{ m}^2/\text{nt})(10^3 \text{ nt/m}^2)(5 \text{ m}^3)$
   $= 3.20 \times 10^{-6} \text{ m}^3.$

3. $\phi = 10^{-2}$;
   thus, $F = S\phi A$
   $= (0.36 \times 10^{12} \text{ dy/cm}^2)(10^{-2})(5 \text{ cm}^2)$
   $= 1.8 \times 10^{10} \text{ dy}.$

4. a $\quad T = \dfrac{PV}{nR} = \dfrac{(10^5 \text{ nt/m}^2)(2 \text{ m}^3)}{(3 \text{ moles})(8.3 \text{ joule/mole °K})}$

   $= 8.0 \times 10^3 \text{ °K}.$

   b $\quad \dfrac{P_1 V_1}{T_1} = \dfrac{P_2 V_2}{T_2}$. Since $T = $ const, $T_1 = T_2$,

and $\quad\quad P_2 = \dfrac{P_1 V_1}{V_2} = \dfrac{(10^5 \text{ nt/m}^2)(2 \text{ m}^3)}{1 \text{ m}^3}$

   $= 2 \times 10^5 \text{ nt/m}^2.$

   c $\quad$ Using the formula in (b), since $P = $ const, $P_1 = P_2$, and
$T = $ const, $T_1 = T_2$; thus,

$V_2 = \dfrac{T_2 V_1}{T_1} = \dfrac{2T_1(2 \text{ m}^3)}{T_1} = 4 \text{ m}^3;$

hence, the volume is doubled.

Answers

0.6 cm	1
$3.20 \times 10^{-6} \text{ m}^3$	2
$1.8 \times 10^{10} \text{ dy}$	3
$8.0 \times 10^3 \text{ °K}$	4 a
$2 \times 10^5 \text{ nt/m}^2$	b
The volume is doubled.	c

in length L − ΔL per unit length $\Delta L/L$ is the corresponding strain. Again, Young's modulus is the proportionality constant and Eq. 10-1 holds.

The **shear stress** is the tangential force per unit area F_T/A. This stress tends to distort the body. The angle ϕ (Fig. 10-4) is the corresponding strain. The **shear modulus** S is the proportionality constant.

$$\frac{F_T}{A} = S\phi \qquad (10\text{-}2)$$

The normal force on a surface per unit area is the **pressure** P; thus, $P = F_N/A$. The relative change in volume $\Delta V/V$ is the corresponding strain. The proportionality constant is $-B$, where,

$$P = \frac{-B\Delta V}{V} \qquad (10\text{-}3)$$

The three moduli Y, S, and B are different for different materials. They depend on the ability of the material to resist distortion due to outside forces. Solids can resist tension and shear stresses, but fluids (ideally) cannot. The **compressibility** k is defined as,

$$k = \frac{1}{B}. \qquad (10\text{-}4)$$

The thermodynamic properties of every pure substance of mass m are related to one another by an equation of state. The temperature T, the volume V, and the pressure P of a mass are the properties considered here. The simplest equation of state is that of an ideal gas,

$$PV = nRT, \qquad (10\text{-}5)$$

where n is the number of moles (1 mole = 6.02×10^{23} molecules) and R is the universal gas constant: $R = 8.315 \times 10^7$ (dy/cm^2)cm^3/mole °K = 8.315×10^7 erg/mole °K = 8.315 joule/mole °K = 1.99 cal/mole °K = 0.082 (liter atmosphere)/mole °K.

The liquid and gas phases can exist together only if the temperature T of a substance is lower than the **critical temperature** T_c. If $T > T_c$, the substance *cannot* be made a liquid by compression.

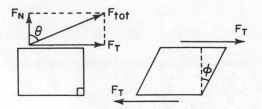

A rectangular solid is distorted into a parallelogram by the shear stress. ϕ is the corresponding strain.

Fig. 10-4

11 TEMPERATURE, CALORIMETRY, THERMAL EXPANSION, AND HEAT TRANSFER

1. A metal rod 1 m long increases its length by 2.00 mm when its heat is increased from 0° C to 100° C. What is the coefficient of linear expansion for this metal?

2. A brass rod is exactly 2.00 m long at 30° C. What is the increase in its length when the rod is heated to 100° C? (The coefficient of linear expansion $\alpha = 19 \times 10^{-6}/$°C for brass.)

3. The length of a column of a certain liquid in a fine capillary tube does not change with an increase in temperature ΔT. If the coefficient of volume expansion $\beta = 10^{-5}/$°C find the α for the material in the tube.

4. A 100 cm³ glass bottle is filled with mercury at 20° C. If the bottle is heated to 100° C, how much mercury will spill over? ($\beta = 18.2 \times 10^{-5}/$°C for mercury, and $\alpha = 9.1 \times 10^{-6}$ for glass.)

5. A thermometer at 5° C is placed in 200 cm³ of water at 45° C. The mass and specific heat of the thermometer are 75 gm and 0.4 cal/gm, respectively. What is the final temperature read on the thermometer?

6. The area of a light bulb's filament is 0.4 cm², its temperature is 3000° K, and the emissivity $e = 0.33$.
   a Calculate the rate at which energy is radiated from the bulb's filament.
   b If the filament's temperature is halved, by what factor is this rate diminished?

7. A glass window separates an ice bath and a boiling water bath. The area of the window is 25 cm² and the thickness of the window is 0.5 cm. ($k = 2.0 \times 10^{-3}$ cal/cm sec °C.)
   a How much heat is transferred by conduction across the window each second?
   b If the wall is replaced by a silver wall, how much heat is transferred across this wall? ($k = 1.0$ cal/cm sec °C.)

1	
2	
3	
4	
5	
6 a	
b	
7 a	
b	

BASIC FACTS

There are three commonly used temperature scales: the **Celsius** (*centigrade*), the **Kelvin** (*ideal gas* or *Absolute*), and the **Fahrenheit** scales. These scales will be designated as T_C, T_K, and T_F, respectively: $T_C = T_K - 273.15°$ K, and $T_C = 5/9 (T_F - 32°$ F$)$.

A solid of length L at Temperature T expands by an amount ΔL if the temperature is increased by an amount ΔT. ΔL is proportional to ΔT; thus,

$$\Delta L = \alpha L \Delta T, \qquad (11\text{-}1)$$

where α is the **coefficient of linear expansion**. The change of area of a solid is,

$$\Delta A = 2\alpha A \Delta T, \qquad (11\text{-}2)$$

and the change of volume of a solid is,

$$\Delta V = 3\alpha V \Delta T. \qquad (11\text{-}3)$$

A liquid of volume V expands an amount ΔV; hence,

$$\Delta V = \beta V \Delta T, \qquad (11\text{-}4)$$

β is the **coefficient of volume expansion**.

If two bodies are placed in contact so that heat can flow from one to the other, they will come to a state of **thermal equilibrium** (the same temperature). Over a limited temperature range, the increase in the temperature of a body is proportional to the heat energy added to that body; thus,

$$\Delta Q = C \Delta T, \qquad (11\text{-}5)$$

where C is the **heat capacity** and ΔQ is the amount of heat added. The heat capacity per unit mass is the **specific heat** c, which is given by,

$$c = \frac{C}{m}, \qquad (11\text{-}6)$$

where m is the mass. **Molar heat capacity** C_m is defined as,

$$C_m = \frac{C}{n}, \qquad (11\text{-}7)$$

(Continued on page 44)

ADDITIONAL INFORMATION

Notice that the length, area, and volume all expand *linearly* with temperature change (Eq. 11-1, 11-2, 11-3, 11-4). When an object is heated, all dimensions expand by the same factor. Areas of holes in objects and the volumes of containers also increase linearly with temperature.

Convection can be forced, as when a fan blows air first through heating coils and then into a room or as the result of convection currents. Convection currents arise because hot air (or liquid) is less dense than cold air, therefore it can be displaced upward (from Archimedes' principle), and cold air moves in to take its place. Convective heating of an object occurs because air molecules lose some of their kinetic energy to the cooler object.

Example 1. Suppose one end of a copper rod is welded to a water bath at 20°C and the other end is welded to a steam pipe at 100° C. The rod is 2 m long and has a cross-sectional area of 3 cm². How much heat flows from the hot end of the rod to the cool end in 1 minute? The conductivity $k = 9.2 \times 10^{-2}$ kcal/sec m for copper; thus, using Eq. 11-8,

$$\begin{aligned}
\Delta Q &= -kA \frac{\Delta T}{\Delta x} \Delta t \\
&= -(9.2 \times 10^{-2}\,\text{kcal/sec m °C})(3 \times 10^{-4}\,\text{m}) \\
&\quad \left(\frac{-80°\,\text{C}}{2\,\text{m}}\right)(60\,\text{sec}) \\
&= 6.62 \times 10^{-2}\,\text{kcal}.
\end{aligned}$$

Notice that $\Delta T = 20° - 100°\,\text{C} = -80°\,\text{C}$.

A wall made of a material of large k is called a **diathermal wall**. Such a wall permits both of its sides to reach thermal equilibrium. If the conductivity for the wall is very small, the material is called an **insulator**. Cork and styrofoam are good examples of insulators. If the conductivity $k = 0$, the wall is called an **adiabatic wall**. Two substances separated by an adiabatic wall are not in thermal contact. They may, however, be in thermal equilibrium with one another, if they happen independently to be at the same temperature or if both are in equilibrium with a third substance.

Radiation is the transfer of energy (heat) by electromagnetic waves. In Eq. 11-9, if the emissivity

(Continued on page 44)

EXPLANATIONS

1. $\alpha = \dfrac{\Delta L}{L \Delta T} = \dfrac{2 \times 10^{-3}\,\text{m}}{(1\,\text{m})(100^\circ\,\text{C})} = 2 \times 10^{-5}/^\circ\text{C}.$

2. $\Delta L = \alpha L \Delta T = (19 \times 10^{-6}/^\circ\text{C})(2\,\text{m})(70^\circ\text{C}) = 2.66 \times 10^{-3}\,\text{m}.$

3. The volume of the capillary tube must increase by the same amount that the volume of liquid increases.
   $\Delta V_{\text{liq}} = 10^{-5} V \Delta T = V_{\text{cap}} = 3 \alpha V \Delta T$; thus,
   $\quad \alpha = 0.33 \times 10^{-5}/^\circ\text{C}.$
   (liq = liquid; cap = capillary)

4. For the bottle, $\Delta V_{\text{bot}} = 3 \alpha V \Delta T = 3(9.1 \times 10^{-6}/^\circ\text{C})(100\,\text{cm}^3)(80^\circ\,\text{C}) = 0.22\,\text{cm}^3$ and, for the mercury, $\Delta V_{\text{merc}} = \beta V \Delta T = (18.2 \times 10^{-5}/^\circ\text{C})(100\,\text{cm}^3)(80^\circ\,\text{C}) = 1.46\,\text{cm}^3$; hence, $(1.46 - 0.22)\,\text{cm}^3 = 1.24\,\text{cm}^3$ will spill out.

5. $\sum\limits_{i=1}^{2} m_i c_i \Delta T_i = 0$; therefore, $(75\ \text{gm})(0.4\ \text{cal/gm}\ ^\circ\text{C})(5^\circ\text{C} - T_f) + (200\ \text{gm})(1\ \text{cal/gm}\ ^\circ\text{C})(45^\circ\text{C} - T_f) = 0$; hence, $9150\ \text{cal} - 230\ \text{cal}\ T_f = 0$ and $T_f = 39.8^\circ\,\text{C}$. The process of measuring T_f actually changes the temperature of the water.

6. a $R' = e\sigma A T^4 = (0.33)(5.67 \times 10^{-5}\ \text{erg/sec cm}^2\ ^\circ\text{K}^4)(0.4\ \text{cm}^2)(3 \times 10^3\ ^\circ\text{K})^4$
   $\qquad = 60.6 \times 10^7\ \text{erg/sec}.$

   b Since the temperature is halved, $T/2 = 1.5 \times 10^3$, $R' = 3.8 \times 10^7\ \text{erg/sec}$ and R' is diminished by a factor of approximately $1/16$, since $(60.6 \times 10^7\ \text{erg/sec})(1/16) = 3.8 \times 10^7\ \text{erg/sec}.$

7. a $\dfrac{\Delta Q}{\Delta t} = -\dfrac{kA\Delta T}{\Delta x}$

   $\qquad = -(2 \times 10^{-3}\ \text{cal/cm sec}\ ^\circ\text{C})(25\ \text{cm}^2)\left(\dfrac{100^\circ\,\text{C}}{0.5\,\text{cm}}\right)$

   $\qquad = -10\ \text{cal/sec}$

   b k is 5×10^2 larger for silver, so that $\dfrac{\Delta Q}{\Delta t}$
   $\qquad = 5000\ \text{cal/sec}.$

Answers

$2 \times 10^{-5}/^\circ\text{C}$	1
$2.66 \times 10^{-3}\,\text{m}$	2
$0.33 \times 10^{-5}/^\circ\text{C}$	3
$1.24\,\text{cm}^3$	4
$39.8^\circ\,\text{C}$	5
$60.6 \times 10^7\ \text{erg/sec}$	6 a
1/16 (approx)	b
$-10\ \text{cal/sec}$	7 a
5000 cal/sec	b

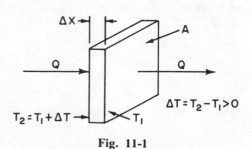

Fig. 11-1

where n is the number of moles, and $n = m/M$, M being the *molecular weight* of the substance.

Convection is the movement of heat resulting from the bulk movement of fluids. If two bodies at unequal temperatures are put together, heat will flow from the hotter to the cooler body by **conduction**. Similarly, if two parts of a body are at different temperatures, heat will flow from the hotter to the cooler region. The slab of material shown in Fig. 11-1 has a cross-sectional area A and thickness Δx. The two faces are kept at temperatures T_1 and $T_2 = T_1 + \Delta T$. The amount of heat ΔQ that flows through the slab per unit time Δt is,

$$\frac{\Delta Q}{\Delta t} = -kA \frac{\Delta T}{\Delta x}, \qquad (11\text{-}8)$$

where k is the **thermal conductivity**. Radiation requires no material medium.

Heat is a form of energy and, therefore, may be expressed in joules, ergs, or foot-pounds, but the following units are conventional: The calorie (cal) is the heat required to raise 1 gram of water from 14.5°C to 15.5°C, 1 kilcalorie (kcal) = 1000 cal, and 1 BTU (British thermal unit) is the heat required to raise 1 standard pound (1/32 slug) of water from 63° F to 64° F. 1 BTU = 252 cal = 0.252 kcal; 1 cal = 4.19 joule.

$e = 1$ for a body, it is called a **black body**. Such a body absorbs all radiation incident upon it; it is both a good absorber and a good radiator of electromagnetic waves. The emissivity for a well polished surface is very small; if the emissivity $e = 0$ the body is called a **perfect reflector**.

When n substances at different temperatures are placed together so that heat can flow among them, they eventually come to the same final temperature T_f. If no heat is lost to the outside; that is, if the substances are isolated adiabatically from the outside.

$$\sum_{i=1}^{n} \Delta Q_i = 0, \qquad (11\text{-}10)$$

where $\Delta Q_i = c_i m_i (\Delta T_i - T_f)$, c_i being the specific heat.

Example 2. Suppose a 1000 gm copper cup at 20° C surrounded by an adiabatic wall is filled with 300 gm of water at 20° and a 500 gm block of aluminum at 60° C is placed in the cup. Find the final temperature. The specific heat of copper, water, and aluminum is 0.093 cal/gm, 1.00 cal/gm, and 0.217 cal/gm, respectively. (Cal/gm is in °C.) Labeling the cup, water, and block as 1, 2, and 3, respectively,

$$\sum_{i=1}^{3} \Delta Q_i = \sum_{i=1}^{3} m_i c_i \Delta T_i = 0;$$

thus (1000 gm)(0.093 cal/gm °C)(20° − T_f) + (300 gm)(1.00 cal/gm °C)(20° − T_f) + (500 gm)(0.217 cal/gm °C)(60° − T_f) = 14,370 − 501.5 T_f = 0; thus $T_f = 28.6$°C.

The rate R at which energy is radiated from a *unit* area of a body per unit time is,

$$R = e\sigma T_K^4, \qquad (11\text{-}9)$$

where e is the *emissivity* and $\sigma = 5.670 \times 10^{-5}$ erg/(sec cm^2 °K^4). Eq. 11-9 is the **Stefan-Boltzmann law of radiation**. If the area is not a unit area, $R' = RA = Ae\sigma T_K^4$.

12

THE LAWS OF THERMODYNAMICS

1. An engine works between 27° C and −23° C. What is the maximum possible efficiency of this engine?

2. A Carnot engine contains 1 mole of an ideal gas for which $\gamma = 1.5$. The upper and lower temperatures are 400° K and 200° K, respectively, $P_1 = 2$ atm, and $P_2 = 1$ atm.
   a How much work is done in process 1? (See Additional Information and Fig. 12-1.)
   b How much heat enters or leaves the system during process 1?
   c What is the change in the internal energy of the gas during process 1?
   d What is the entropy change of the gas during process 1?
   e How much heat enters or leaves the system during process 2?
   f What is the change in internal energy for process 2?

g What is the work done during process 3?
h What is the work done and what is the internal energy change during process 4? (Use the results of f and g.)

3. Use the results obtained in problem 1 to calculate the following:
   a The change in internal energy of the gas for the entire cycle.
   b The work for the cycle.

4. Twenty grams of ice melts into water at 0° C.
   a What is the entropy change of the ice?
   b Is the water in a more or less orderly state in this liquid form?
   c If the water is refrozen, what is the entropy change of the water?
   d What is the total entropy change of the water in melting and refreezing?

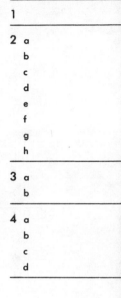

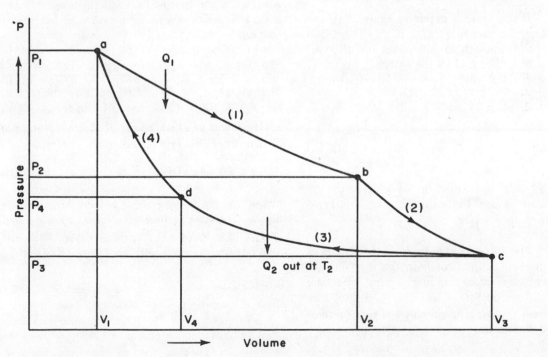

Fig. 12-1

BASIC FACTS

A *system* is a body which can interact with its environment. The environment may work on or be worked on by the system. A *process* is any activity of the system, such as undergoing compression or absorbing heat.

Both heat and work are forms of energy, but heat can only be transferred as a result of temperature difference. Heat is therefore energy *in transit* from a hotter to a cooler body. Work, on the other hand, is energy which is transferred by some force F causing a displacement dx; thus $dW = Fdx$. (See chapter 4.)

A change in the amount of heat Q flowing into a system is designated by ΔQ; a small amount of heat, or differential of heat, is dQ. If they are added to the system under consideration, ΔQ and dQ are positive. A small amount of work dW which a system performs on its environment is positive. If the environment does work on the system, $dW < 0$.

If the system expands against an external pressure P, and if the expansion is slow enough so the pressure of the system is the same as the external pressure; that is, if the expansion is *quasi-static*, then,

$$dW = P\,dV \qquad (12\text{-}1)$$

and, $$W = \int_{V_1}^{V_2} P\,dV \qquad (12\text{-}1a)$$

Since P is a function of V, it must be known how the expansion or compression took place.

The **internal energy** U of a system is defined as the *change* in the internal energy ΔU occurring in a system, when the system performs work W and absorbs heat Q; therefore, the **first law of thermodynamics** is,

$$\Delta U = U_f - U_i = Q - W, \quad (12\text{-}2)$$

(Continued on page 48)

ADDITIONAL INFORMATION

The first law of thermodynamics is a statement of the conservation of energy. The fact that the internal energy does not depend on the path is very important. Work can be shown to depend on the path for a cylinder of gas fitted with a piston. In this case, $dW = PdV$. The total work performed when the volume goes from V_1 to V_2 is given by Eq. 12-1a. The heat Q absorbed by two expanding gases a and b is different and the difference $Q_a - W_a = Q_b - W_b$ is equal ($W =$ work). This difference is the internal energy change ΔU of the gas, which depends only on its initial and final states. If heat is given off by a system, $Q < 0$; if work is performed on the system, $W < 0$. If the only work possible is dW, then,

$$dQ = dU + PdV. \text{ (See Eq. 12-1 and 12-2a.)} \quad (12\text{-}4)$$

At constant volume ($dV = 0$) and,

$$dQ = nC_v dT = dU. \text{ (See Eq. 12-2a)} \quad (12\text{-}4a)$$

Here, C_v is the specific heat at a constant volume. At constant pressure,

$$dQ = nC_p dT, \qquad (12\text{-}4b)$$

where C_p is the specific heat at a constant pressure. Since U depends only on temperature (is independent of P and V),

$$dU = nC_v dT. \qquad (12\text{-}5)$$

Thus $nC_p dT = nC_v dT - PdV$, but from the ideal gas law, $pdV = nRdT$ if $P =$ const. Upon combining Eq. 12-5a and 12-5b, $C_p - C_v = R = 8.32$ joule/mole °K = 2 cal/mole °K.

For a gas undergoing an adiabatic process ($dQ = 0$), $PV^\gamma = K$, where $\gamma = C_p/C_v$. The constant K depends on the amount of gas and on the initial conditions. The work performed by a system when its volume changes by an amount dV is $dW = PdV$. For an ideal gas, $PV = nRT$, so that $dW = nRTdV/V$.

The **Carnot cycle** is a series of reversible processes. Two of the processes are isothermal and operate at temperatures T_1 and T_2. The system, a quantity of ideal gas, takes in an amount of heat Q_1 when it is operating at T_1 and gives up Q_2 when operating at T_2.

(Continued on page 48)

EXPLANATIONS

1. $e = \dfrac{T_1 - T_2}{T_1}$; $T_1 = (27 + 273)° \text{ K} = 300 ° \text{K}$ and $T_2 = (-23 + 273)° \text{ K} = 250° \text{K}$. Thus $e = \dfrac{300° - 250°}{300°} = 0.167$.

2. a From the ideal gas law, $PV = nRT = \text{const.}$ Thus $V_2 = 2V_1$, since $P_2 = \frac{1}{2} P_1$ and $T_1 = \text{const.}$ $W_1 = nRT_1 \ln \dfrac{V_2}{V_1} = (1 \text{ mole})(8.32 \text{ joule/mole } °\text{K})(400° \text{ K})$ $(\ln 2) = (3328 \text{ joule})(0.693) = 2306 \text{ joule}$.

   b $\Delta Q_1 = W_1 = 2306 \text{ joule}$, since $\Delta U_1 = 0$.

   c $\Delta U = 0$, since U depends on T only.

   d $\Delta S = \dfrac{\Delta Q_1}{T_1} = \dfrac{2306 \text{ joule}}{400° \text{ K}} = 5.77 \text{ joule/}°\text{K}$.

   e $\Delta Q_2 = 0$, since process 2 is an adiabatic process.

   f $\Delta U = 0$.

   g $W_3 = nRT_2 \ln \dfrac{V_4}{V_3}$. We must calculate V_4. First we find the equation for the adiabatic ad: $P_1 = 2 \text{ atm} = 2 \times 10^5 \text{ nt/m}^2$. $K_2 = P_1 V^\gamma = (2 \times 10^5 \text{ nt/m}^2)$ $(0.0167 \text{ m}^3)^{3/2} = (2 \times 10^5 \text{ nt/m}^2)(2.16 \times 10^{-3} \text{ m}^3) = 4.32 \times 10^2$, thus $K_2 = 432$ (no units). Furthermore, $p_4 V_4 = nRT_2$, therefore, $V_4^{\gamma-1} = V_4^{1/2} = \dfrac{432}{(1 \text{ mole})(8.32 \text{ joule/mole } °\text{K})(200° \text{ K})} = 0.259 \text{ m}^3$, and $V_4 = 0.066 \text{ m}^3$. Hence, $W_3 = nRT \ln \dfrac{V_4}{V_3} = 1155 \text{ joule}$.

   h $W_4 = -W_2 = -1715 \text{ joule}$; $\Delta U_4 = -W_4 = +1715 \text{ joule}$.

3. a $\Delta U = \Delta U_1 + \Delta U_2 + \Delta U_3 + \Delta U_4 = 0 + \Delta U_2 + 0 + (-U_2) = 0$.

   b $W = W_1 + W_2 + W_3 + W_4 = W_1 + W_3 = 2306 - 1155 = 1151 \text{ joule}$. (Remember that $W_2 = -W_4$.)

4. a $\Delta S = \dfrac{\Delta Q}{T} = \dfrac{(20 \text{ gm})(80 \text{ cal/gm})}{273° \text{ K}} = 05.86 \text{ cal/}°\text{K}$.

   b ΔS is positive; thus, the liquid phase is much less orderly than the solid phase.

   c $\Delta S = \dfrac{\Delta Q}{T} = - 05.86 \text{ cal/}°\text{K}$.

   d $\Delta S = 0$. Since the melting-refreezing cycle is reversible, ΔS for the complete cycle must be zero.

Answers

0.167	1
2306 joule	2 a
2306 joule	b
0	c
5.77 joule/°K	d
0	e
0	f
1155 joule	g
−1715 joule; +1715 joule	h
0	3 a
1151 joule	b
05.86 cal/°K	4 a
Less orderly	b
05.86 cal/°K	c
0	d

or, $dU = dQ - dW,$ (12-2a)

where the subscripts f and i stand for the *final* and *initial* states, respectively.

A process, such as the compression or expansion of a gas, is *quasi-static* if it occurs so slowly that the pressure can be plotted for every value of volume of the gas. If the process is too rapid, the pressure is not uniform in the gas and a plot is not possible. In a quasi-static process, a system is at equilibrium at every instant. A process is *reversible* if it is able to retrace its original path. All reversible processes are quasi-static, but not all quasi-static processes are reversible. The presence of friction in a piston which compresses a gas is enough to make the process irreversible.

Entropy ΔS, which is a measure of *disorder*, is given by,

$$\Delta S = \frac{Q}{T},$$ (12-3)

where $\Delta S = S_f - S_i$, the "final" and "initial" equilibrium states and Q is the heat absorbed by the system. If T is not constant, then,

$$\Delta S = \int_{S_i}^{S_f} \frac{dQ}{T}$$ (12-3a)

The Second Law of Thermodynamics.
1. **Clausius' statement:** *It is impossible for a cyclic engine to conduct heat from a cold body to a hot body without doing work.*
2. **Kelvin's statement:** *It is impossible to construct a cyclic heat engine with efficiency e = 1.*

The efficiency e of the engine is,

$$e = \frac{T_1 - T_2}{T_1}.$$ (12-6)

Process 1.   The gas is initially in an equilibrium state (P_1, V_1, T_1) at point a in Fig. 12-1. The gas expands isothermally and reversibly to a new state (P_2, V_2, T_1) at point b in the figure. (Along curve ab, $T = T_1$.)

Process 2.   The gas expands adiabatically and reversibly to (P_3, V_3, T_2) at c.

Process 3.   The gas is compressed isothermally to (P_4, V_4, T_2) at d.

Process 4.   The gas is compressed adiabatically to its original state (P_1, V_1, T_1). (Table A-1 in the Appendix shows the states of the system during its processes.)

Energy is conserved in any *adiabatic* system. A block sliding along a horizontal, waxed surface eventually slows down because of friction. The kinetic energy of motion is not destroyed, however it is changed into heat energy. This process involves a change from an *orderly* form of kinetic energy where, for example, every molecule of the block has the same velocity in the same direction, to a *disordered* form of kinetic energy with random molecular motion within the block and on its surface.

The efficiency e is given by,

$$e = \frac{\text{output work}}{\text{heat taken in at high temperature}}$$ (12-7)

$$= \frac{W}{Q_1} = \frac{Q_1 - Q_2}{Q_1} = 1 - \frac{Q_2}{Q_1};$$

also, see Eq. 12-6.

13

1. The speed of sound in air on a certain day is 340 m/sec. What is the wavelength of a 1700 Hertz (cycles per second) sound wave?

2. The bulk modulus for water is 2.1×10^{10} dy/cm^2 and the density of water is 1 gm/cm^3. Find the velocity of sound in water.

3. A certain wire is 2 m long and has a mass of 0.04 kg. It is stretched with a tension of 2 nt.
   a Find the wave velocity of transverse waves in the wire.
   b Find the fundamental frequency (lowest frequency) of the wire.
   c If the wire is now stretched with a tension of 6 nt, what is the new velocity of transverse waves on the wire.

4. A man watching a fireworks display sees an explosion 4 sec before he hears the sound. If the velocity of sound in air is 1100 ft/sec, how far is the man from the explosion?

5. Two tuning forks which vibrate at 256 Hertz and at 260 Hertz are struck at the same time. Give the resulting frequency and the beat frequency.

6. A siren on a police car has a frequency of 1200 Hertz.
   a If the police car is approaching a stationary observer at 30 m/sec, what frequency does the observer hear? (Take the velocity of sound in air to be 330 m/sec.)
   b If the observer is escaping the police car in another car moving at 40 m/sec, what frequency does he hear?

7. A brass pipe and a steel pipe both 100 m long are struck at one end; the sound waves produced travel the length of the pipes. What is the time interval between the arrival of the two waves at the other end of the pipes? (Young's modulus for steel and brass are 2.0×10^{12} dy/cm^2 and 0.91×10^{12} dy/cm^2, and the densities of steel and brass are 7.8 gm/cm^3 and 8.6 gm/cm^3, respectively.)

1 _____

2 _____

3 a _____
  b
  c _____

4 _____

5 _____

6 a _____
  b _____

7 _____

BASIC FACTS

A **mechanical wave** is a disturbance which moves through a *deformable* medium, which might be a solid, liquid, or gas. When a mechanical wave passes through a medium, each portion of that medium *oscillates* about its equilibrium position (the position it would hold if no waves were present).

The equation for the amplitude of the displacement y for a **sine wave** is,

$$y = y_0 \sin 2\pi \left(\frac{x}{\lambda} \pm \frac{t}{T} \right) \qquad (13\text{-}1)$$

$+$ or $-$ indicates that the wave is moving to the *right* or *left*, respectively, λ is the *wavelength,* and T is the *period*. The *velocity* v of the wave is,

$$v = \lambda f, \qquad (13\text{-}2)$$

where f, the *frequency* of the wave in Hertz (cycles/sec), is related to the *radian frequency* ω by

$$f = \frac{\omega}{2\pi}, \qquad (13\text{-}3)$$

and to T by $\qquad f = \frac{1}{T}. \qquad (13\text{-}3a)$

The **wave number** k is given by,

$$k = \frac{2\pi}{\lambda} \qquad (13\text{-}4)$$

Using k and ω, Eq. 13-1 becomes $y = y_0 \sin(kx \pm \omega t + \phi)$, where ϕ is the **phase constant.**

The velocity of a wave depends on the medium and the type of wave. **Transverse waves** in a piece of string have a velocity,

$$v = \sqrt{\frac{F}{\mu}}, \qquad (13\text{-}5)$$

where F is the tension and μ is the linear density (mass per unit length). **Longitudinal waves** in a fluid have velocity,

(Continued on page 52)

ADDITIONAL INFORMATION

Mechanical waves occur in all matter and are of several types. If the matter oscillates *perpendicularly* to the direction in which the wave moves, the wave is called a **transverse wave**; if the matter oscillates in the *same direction* as the wave motion, the wave is called a **longitudinal wave**. The waves on a piece of string are transverse, sound waves are longitudinal, and surface water waves are a mixture of both.

The equation for a wave generally depends on the expression $x \pm vt$; the most important waveform is,

$$y = y_0 \sin \frac{2\pi}{\lambda} (x \pm vt), \qquad (13\text{-}7)$$

where the $(+)$ sign is used if a wave is traveling to the *left* and the $(-)$ sign is used if the wave is traveling to the *right*. The sine curve in Fig. 13-1 moves to the right; $y = 0$ at $x = 0$ and $t = 0$, but at some later time, the wave has shifted so that when $y = 0$, $x \neq 0$, thus $x - vt = 0$, or $v = x/t$.

Standing waves can be produced by plucking a stretched string which is firmly clamped at both ends. Two waves are created which travel in opposite directions; according to the superposition principle,

$$y_1 = y_0 \sin(kx - \omega t)$$

and, $\qquad y_2 = y_0 \sin(kx + \omega t)$

represent two waves traveling toward the right and left, respectively. The total amplitude is just the sum of y_1 and y_2; thus, $y = y_1 + y_2 = 2y_0 \sin kx \cos \omega t$, using the trigonometric identity $\sin A + \sin B = 2 \sin \frac{1}{2}(A + B) \cos \frac{1}{2}(A - B)$, where $A = kx + \omega t$ and $B = kx - \omega t$. Because the endpoints of the string cannot move, $y = 0$ where $x = L = 0$ (L is the length) or when $k = 0$; that is, the wave must have a node (zero amplitude) at the endpoint. Thus only certain values of λ can occur on the string and $y = 2y_0 \sin 0 \cos \omega t = 0$ for all time; also, $y = 2y_0 \sin kL \cos \omega t = 0$ only if $\sin kL = 0$. This means that $kL = 2\pi L/\lambda = N\pi$; thus, $\lambda = 2L/N$, where N is any integer.

The frequencies which can occur in an organ pipe open at one end are,

$$f = \frac{nv}{2L}; \qquad (n = 1, 2, 3, \ldots)$$

(Continued on page 52)

EXPLANATIONS

1. $v = \lambda f$; thus, $\lambda = \dfrac{v}{f} = \dfrac{340 \text{ m/sec}}{1700 \text{ Hertz}} = 0.20 \text{ m} = 20 \text{ cm}.$

2. $v = \sqrt{\dfrac{B}{\rho}} = \sqrt{\dfrac{2.1 \times 10^{10} \text{ dy/cm}^2}{1 \text{ gm/cm}^3}} = 1.45 \times 10^5 \text{ cm/sec}.$

3. a $v = \sqrt{\dfrac{T}{\mu}}$ and $\mu = \dfrac{0.04 \text{ kg}}{2 \text{ m}} = 0.02 \text{ kg/m}$; thus,

   $v = \sqrt{\dfrac{2 \text{ nt}}{0.02 \text{ kg/m}}} = 10 \text{ m/sec}.$

   b $\lambda = \dfrac{L}{2} = 1 \text{ m}$; $f = \dfrac{v}{\lambda} = \dfrac{10 \text{ m/sec}}{1 \text{ m}} = 10 \text{ Hertz}.$

   c T is increased by a factor of 3; thus, $v' = \sqrt{3}\, v = \sqrt{3}\,(10 \text{ m/sec}) = 17.3 \text{ m/sec}$ is the new velocity.

4. $v = \dfrac{x}{t}$; thus $x = vt = (1100 \text{ ft/sec})(4 \text{ sec}) = 4400 \text{ ft}.$

5. The resulting frequency is $\frac{1}{2}(260 + 256)$ Hertz = 258 Hertz; the beat frequency is $\frac{1}{2}(260 - 256)$ Hertz = 2 Hertz.

6. a $f' = f\left(\dfrac{v}{v - v_s}\right) = (1200 \text{ Hertz})\dfrac{330 \text{ m/sec}}{(330 - 30) \text{ m/sec}}$

   $= 1.32 \times 10^3 \text{ Hertz}.$

   b $f' = f\left(\dfrac{v + v_0}{v - v_s}\right) = (1200 \text{ Hertz})\dfrac{(330 - 40) \text{ m/sec}}{(330 - 30) \text{ m/sec}}$

   $= 1160 \text{ Hertz}.$

   From the Doppler shift, the driver can tell whether or not the police car is gaining on him.

7. For steel $v_1 = \sqrt{\dfrac{Y_1}{\rho_1}} = \sqrt{\dfrac{2.0 \times 10^{12} \text{ dy/cm}^2}{7.8 \text{ gm/cm}^3}}$

   $= 5.1 \times 10^5 \text{ m/sec};$

Answers

20 cm	1
1.45×10^5 cm/sec	2
10 m/sec	3 a
10 Hz	b
17.3 m/sec	c
4400 ft	4
Resulting: 258 Hz	5
Beat: 2 Hz	
1.32×10^3 Hz	6 a
1160 Hz	b
0.93×10^{-4} sec	7

for brass, $v_2 = \sqrt{\dfrac{Y_2}{\rho_2}}$

$= \sqrt{\dfrac{0.91 \times 10^{12} \text{ dy/cm}^2}{8.6 \text{ gm/cm}^3}}$

$= 3.46 \times 10^5 \text{ m/sec}.$

The time required for the wave to reach the end of the steel rod is $t_1 = \dfrac{x}{v_1} = \dfrac{100 \text{ m}}{5.1 \times 10^5 \text{ m/sec}} = 1.96 \times 10^{-4}$ sec; the time for the wave to reach the end of the brass rod is $t_2 = \dfrac{x}{v_2} = \dfrac{100 \text{ m}}{3.46 \times 10^5 \text{ m/sec}} = 2.89 \times 10^{-4}$ sec; thus, $t_2 - t_1 = 0.93 \times 10^{-4}$ sec.

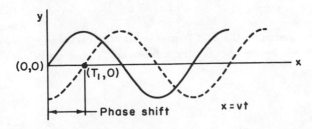

Fig. 13-1

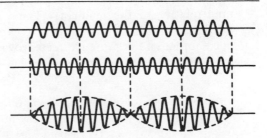

Fig. 13-2

$$v = \sqrt{\frac{B}{\rho}}, \qquad (13\text{-}5a)$$

where ρ is the density of the material, and B is the **bulk modulus of elasticity.** Longitudinal waves in a gas have a velocity,

$$v = \sqrt{\frac{\gamma p}{\rho}}, \qquad (13\text{-}5b)$$

where the ratio of specific heats $\gamma = C_p/C_v$ (chapter 12), p is the pressure, and ρ is the density of the gas. Longitudinal waves in a thin rod have velocity,

$$v = \sqrt{\frac{Y}{\rho}}, \qquad (13\text{-}5c)$$

where Y is **Young's modulus** (chapter 10).

Two waves can pass over the same region independently of one another. This is called the **superposition principle,** which implies that if two waves y_1 and y_2 intersect, the total displacement is $y_{\text{total}} = y_1 + y_2$.

If the source of waves is moving with velocity v_s and the observer moves with velocity v_0, then the frequency f' of the wave which the observer measures is different from the frequency f which the source emits; thus,

$$f' = f\left(\frac{v + v_0}{v - v_s}\right), \qquad (13\text{-}6)$$

where v is the velocity of the wave in a stationary medium. The velocities of both the observer and source are measured relative to the medium. The velocities are *positive* if directed toward one another and *negative* if directed away from one another.

for a closed organ pipe,

$$f = \frac{nv}{4L}, \qquad (n = 1, 3, 5, \ldots.)$$

The **Doppler effect** is a shift in the frequency of a wave because the source and/or the observer of the wave are moving relative to one another. If the velocity of the source $v_s = 0$ and the velocity of the observer $v_0 \neq 0$, then,

$$f' = f\left(\frac{v + v_0}{v}\right), \qquad (13\text{-}6a)$$

where v is the wave velocity in the medium. If $v_s \neq 0$ and $v_0 = 0$, then,

$$f' = f\left(\frac{v}{v - v_s}\right). \qquad (13\text{-}6b)$$

If both source and observer are moving (relative to the medium),

$$f' = f\left(\frac{v + v_0}{v - v_s}\right). \qquad (13\text{-}6c)$$

If two waves of slightly different frequencies f_1 and f_2 are traveling in the same direction, **beats** occur and,

$$y_1 = y_0 \sin 2\pi f_1 t$$
and
$$y_2 = y_0 \sin 2\pi f_2 t.$$

The amplitudes of both waves are assumed to be equal and the position of the wave was chosen to be $x = 0$, then using the same trigonometric identity as before,

$$y = y_1 + y_2 = 2y_0 \sin 2\pi\left(\frac{f_1 + f_2}{2}\right)t \cos 2\pi\left(\frac{f_1 - f_2}{2}\right)t,$$

where $2\pi f_1 t = A$ and $2\pi f_2 t = B$. The equation for the amplitude (dotted curve) in Fig. 13-2 is,

$$2y_0 \cos 2\pi\left(\frac{f_1 - f_2}{2}\right)t.$$

Therefore, the **beat frequency** (dotted curve) is $(f_1 - f_2)/2$ and the **resulting frequency** (solid curve) is $(f_1 + f_2)/2$.

14

ELECTROSTATICS

1. If two charges, $q_1 = 5 \times 10^{-3}$ coul and $q_2 = -2 \times 10^{-4}$ coul, are separated by 3 m, what is the force between them?

2. If E at a point is 75 nt/coul, what is the force on a 0.3 coul charge placed at the point?

3. Calculate the electric field for a 10^{-5} coul charge, 4 cm along the axis of a charged ring (in Fig. 14-1, $z = 0.04$ m), where $Q = 2 \times 10^{-5}$ coul, and $a = 0.03$ m.

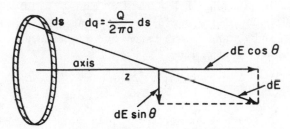

$$dq = \frac{Q}{2\pi a} ds$$

Fig. 14-1

4. a Find the magnitude of the electric field that will exactly balance the weight of an electron, if the electron's mass is 9.11×10^{-31} kg and its charge is 1.6×10^{-19} coul.

   b Find the magnitude of the electric field that will exactly balance the weight of a 1 coul charge of mass 0.001 kg.

5. An electron traveling at 5×10^6 m/sec enters a region in which the electric field is parallel to its motion; after traveling 0.05 m, the velocity of the electron is 3×10^6 m/sec. The mass and charge on the electron are the same as in problem 4a. What is the magnitude of the electric field in the region?

6. Calculate the electric field at a distance r from the axis of a uniformly charged cylinder. The radius of the cylinder is R and the charge per unit length is λ. (Be careful inside the cylinder!)

7. Calculate the electric field at a distance r from the center of a charged spherical shell of radius R, if the charge on the shell is Q.

8. An electric dipole consists of two charges 2×10^{-7} coul and -2×10^{-7} coul separated by 10^{-5} m. The dipole is in an electric field of 500 nt/coul and makes an angle of 30° with the electric field.

   a Calculate the electric dipole moment.

   b Calculate the torque on the dipole.

   c Calculate the potential energy of the dipole.

BASIC FACTS

The force between two charges q_1 and q_2 is given by **Coulomb's law,**

$$\mathbf{F}_{12} = \frac{1}{4\pi\epsilon_0} \frac{q_1 q_2 \mathbf{e}_{12}}{r_{12}^2} = -\mathbf{F}_{21}, \quad (14\text{-}1)$$

where $\mathbf{F}_{12}$ is the force in newtons on q_1, r_{12} is the distance from q_1 to q_2 in meters, and $\mathbf{e}_{12}$ is the unit vector in the direction from q_1 to q_2. The charges are in coulombs, $1/4\pi\epsilon_0 = 9.0 \times 10^{-9}$ nt m^2/coul^2, the **permittivity constant** $\epsilon_0 = 8.854 \times 10^{-12}$ coul^2/nt m^2. If one charge is positive and the other negative, $\mathbf{F}$ is *attractive*; if both charges are positive or both negative, $\mathbf{F}$ is *repulsive*.

A field is a space in which a value can be assigned to every point in that space. The value of the **electric field** E at a point is the force per unit positive charge; thus, the electric field E at a point a a distance r from a charge q is,

$$\mathbf{E} = \frac{1}{4\pi\epsilon_0} \frac{q}{r^2} \mathbf{e}, \quad (14\text{-}2)$$

where $\mathbf{e}$ is directed from the charge to the point. If q is positive, $\mathbf{E}$ is directed away from q. The force $\mathbf{F}$ on a charge q is $\mathbf{F} = q\mathbf{E}$. If there are n charges present, then, the total electric field $\mathbf{E}_{tot} = \mathbf{E}_1 + \mathbf{E}_2 + \cdots + \mathbf{E}_n$, by vector addition. If E represents a *continuous* charge distribution, then each point charge element is represented by the *differential dq* and the vector integral must be used; thus,

$$E = \int d\mathbf{E} = \frac{1}{4\pi\epsilon_0} \int \frac{\mathbf{e}\,dq}{r^2}. \quad (14\text{-}2a)$$

Gauss's law for the flux Φ through a *Gaussian surface S* (a closed surface with no holes) is,

$$\Phi = \mathbf{E} \cdot \mathbf{S} = \frac{q}{\epsilon_0}, \quad (14\text{-}3)$$

if $\mathbf{E}$ makes the same angle with $\mathbf{S}$ and is the same magnitude everywhere on $\mathbf{S}$. If

(Continued on page 56)

ADDITIONAL INFORMATION

Whenever the charge distribution is continuous, Eq. 14-2a must be used. If the direction of E is known, the calculation is easy.

Example 1. In Fig. 14-2a, q_1, q_2, and q_3 are the charges at the corners of a square, the sides of which

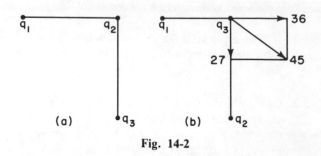

Fig. 14-2

are 10.0 cm long; $q_1 = 4.0 \times 10^{-7}$ coul, $q_2 = 1.0 \times 10^{-7}$ coul, and $q_3 = -3.0 \times 10^{-7}$ coul. Calculate F_{21} and F_{23} and represent them by vectors; calculate F_{net} and E_{net} on q_2 using, $F_{net} = q_2 E_{net}$.

$$\mathbf{F}_{21} = \frac{q_1 q_2 \mathbf{e}_{21}}{4\pi\epsilon_0 r_{21}^2}$$

$$= \frac{(9 \times 10^9 \text{ nt m}^2/\text{coul})(4 \times 10^{-7} \text{ coul})(1 \times 10^{-7} \text{ coul})\mathbf{e}_{21}}{(0.1 \text{ m})^2}$$

$$= (36 \times 10^{-3})\mathbf{e}_{21} \text{ nt};$$

similarly, $\mathbf{F}_{23} = (-27 \times 10^{-3})\mathbf{e}_{23}$ nt. Thus,

$$F_{net} = \sqrt{36^2 + (-27)^2} \times 10^{-3} \text{ nt}$$
$$= (49.2 \times 10^{-3})\mathbf{e}_{23} \text{ nt in the direction}$$

shown in Fig. 14-2b. $E_{net} = F_{net}/q_2 = 49.2 \times 10^4$ nt/coul.

Example 2. Calculate the electric field E at a distance r from a long, charged wire. The charge per unit length is λ coul/m. Choose a Gaussian surface composed of three surfaces S_1, S_2, and S_3 with the same general shape as the charge distribution. Let S_1 be a hollow cylinder of radius r and length L with S_2 and S_3 the circular ends of the cylinder. $\mathbf{E}$ is directed from the positive wire toward the negative wire. In both cases, $\mathbf{E}$ is perpendicular to dS_2 and dS_3, so that $\Phi_2 = \Phi_3 = 0$, but $\mathbf{E}$ is parallel to dS_1. Furthermore, since every point on S_1 is the same distance from the

(Continued on page 56)

EXPLANATIONS

1. $F = \dfrac{q_1 q_2}{4\pi\epsilon_0 r^2}$

   $= \dfrac{(9 \times 10^9 \text{ nt m}^2/\text{coul}^2)(5 \times 10^{-3} \text{ coul})}{(3 \text{ m})^2(-2 \times 10^4 \text{ coul})}$

   $= -10^3 \text{ nt.}$

2. $F = qE = (0.3 \text{ coul})(75 \text{ nt}/\text{coul}) = 22.5 \text{ nt.}$

3. $E = \dfrac{Qz}{4\pi\epsilon_0(z^2 + a^2)^{3/2}}$

   $= \dfrac{(9 \times 10^9 \text{ nt m}^2/\text{coul}^2)(0.04 \text{ m})}{(0.04^2 \text{ m} + 0.03^2 \text{ m})^{3/2}(2 \times 10^5 \text{ coul})}$

   $= 5.76 \times 10^7 \text{ nt}/\text{coul.}$

4. a $mg = qE$, thus $E = \dfrac{mg}{q}$

   $= \dfrac{(9.11 \times 10^{-31} \text{ kg})(9.8 \text{ m}/\text{sec}^2)}{(1.6 \times 10^{-19} \text{ coul})}$

   $= 5.58 \times 10^{-11} \text{ nt}/\text{coul.}$

   b $E = \dfrac{mg}{q} = \dfrac{(0.001 \text{ kg})(9.8 \text{ m}/\text{sec}^2)}{1 \text{ coul}}$

   $= 9.8 \times 10^{-3} \text{ nt}/\text{coul.}$

5. Use $v_1^2 = v_2^2 + 2ax$ (chapter 2) and solve for a; thus,

   $a = \dfrac{v_1^2 - v_2^2}{2x} = \dfrac{(5 \times 10^6 \text{ m}/\text{sec})^2 - (3 \times 10^6 \text{ m}/\text{sec})^2}{2(0.05 \text{ m})}$

   $= 1.6 \times 10^{13} \text{ m}/\text{sec}^2.$

   Then, $E = \dfrac{F}{q} = \dfrac{ma}{q} = \dfrac{(9.11 \times 10^{-31} \text{ kg})(1.6 \times 10^{13} \text{ m}/\text{sec}^2)}{(1.6 \times 10^{-19} \text{ coul})}$

   $= 91.1 \text{ nt}/\text{coul.}$

6. The electric field E is everywhere radial, thus only the curved surface S_1 contributes to Gauss's law. Therefore, $ES_2 = ES_3 = 0$ and $S_1 = 2\pi r L$. The charge q enclosed in S is λL if $r > R$. In this case, Gauss's law is $E(2\pi r L) = \lambda L/\epsilon_0$; hence, $E = \lambda/2\pi\epsilon_0 r$. If $r < R$, the charge enclosed in S is proportional to the volume enclosed. Therefore, $q = \lambda r^2 L/R^2$ and $E(2\pi r L) = \lambda r^2 L/\epsilon_0 R^2$, hence $E = \lambda r/\epsilon_0 R^2$.

7. The Gaussian surface S is a sphere concentric with the charged sphere. The radius of S is r and the area is $4\pi r^2$. The charge enclosed in S is Q if $r > R$; hence, $E(4\pi r^2) = Q/\epsilon_0$ and $E = Q/4\pi\epsilon_0 r^2$. If $r < R$, Q enclosed in S is zero, hence $E = 0$.

8. a $p = qa = (2 \times 10^{-7} \text{ coul})(10^{-5} \text{ m}) = 2 \times 10^{-12} \text{ coul m.}$

Answers

-10^3 nt	1
22.5 nt	2
5.76×10^7 nt/coul	3
5.58×10^{-11} nt/coul	4 a
9.8×10^{-3} nt/coul	b
91.1 nt/coul	5
$\lambda/2\pi\epsilon_0 r \, (r > R),\ \lambda r/\epsilon_0 R^2 \, (r < R)$	6
$Q/4\pi\epsilon_0 r^2 \, (r > R),\ 0 \, (r < R)$	7
2×10^{-12} coul m	8 a
5×10^{-10} nt m	b
-8.66×10^{-10} joule	c

b $\tau = pE \sin \theta = (2 \times 10^{-12} \text{ coul m})(5 \times 10^2 \text{ nt}/\text{coul})(\sin 30°)$

$= (10 \times 10^{-10} \text{ nt m})\left(\tfrac{1}{2}\right) = 5 \times 10^{-10} \text{ nt m.}$

c $U = -pE \cos \theta = -(2 \times 10^{-12} \text{ coul m})(5 \times 10^2 \text{ nt}/\text{coul})(\cos 30°)$

$= -(10 \times 10^{-10} \text{ nt m})(0.866) = -8.66 \times 10^{-10}$ joule.

E and/or the angle θ between **E** and **S** are not constant, the integral formula must be used; thus,

$$\Phi = \int \mathbf{E} \cdot d\mathbf{S} = \frac{q}{\epsilon_0}. \quad \text{(14-3a)}$$

The following important facts are a result of Gauss's law:

1. $E = 0$ within a conductor.
2. The charge on a closed metal can is entirely on the outside of the can.

An **electric dipole** is a pair of charges $+q$ and $-q$ separated by a distance a. The dipole moment is represented by a vector **p** having magnitude $p = qa$ and direction from $-q$ to $+q$.

An electric field exerts a torque τ, which tends to turn **p** so that it lines up with **E**. Here the vector product is used,

$$\tau = \mathbf{p} \times \mathbf{E}, \quad \text{(14-4)}$$

where θ is the angle between **p** and **E**. The magnitude of τ is,

$$\tau = pE \sin \theta \quad \text{(14-4a)}$$

The potential energy U of a dipole in an electric field is,

$$U = -\mathbf{p} \cdot \mathbf{E} = -pE \cos \theta. \quad \text{(14-5)}$$

The unit for electric charge in the mks system is the coulomb (coul): 1 coul is the amount of charge flowing past a given point in a conductor in 1 second when there is a steady current of 1 ampere (amp). The unit of *electric intensity* or *electric field strength* is 1 newton/coulomb (nt/coul).

wire, **E** and θ are constant over the entire surface; therefore, $\Phi = \mathbf{E} \cdot \mathbf{S}_1 = E(2\pi rL) = q/\epsilon_0$. Because q is the charge enclosed inside S, $q = \lambda L$. Thus, $E(2\pi rL) = \lambda L/\epsilon_0$ and $E = \lambda/2\pi\epsilon_0 r$.

15 ELECTRIC POTENTIAL AND CAPACITANCE

1. A -10^{-5} coul charge was originally a distance of 70 cm from a large charged metal plate. It is brought to a position 20 cm from the plate. The charge density σ of the plate is $5 \times 10^{-4} \, \text{coul}/\text{m}^2$.
   a What is the force on the charge at both positions?
   b What is the change in potential of the charge?
   c What is the change in the potential energy of the charge?

2. a Calculate the potential at a point P 1 m from a dipole (Fig. 15-1), where $q = 0.5$ coul, $a = 10^{-6}$ m.
   b Calculate the potential energy of a 1.0 coul charge where $r = 1$ m and $\theta = 60°$.

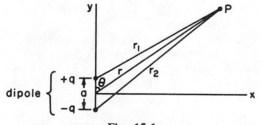

Fig. 15-1

3. Calculate the potential both inside and at a distance r from the center of a spherically charged shell, where the charge on the shell is q, and the radius of the shell is R.

4. Calculate the potential at a distance r from the center of a uniformly charged sphere of radius R and charge q.

5. Two plates, each with an area of 10 m^2, are separated by 1 cm of air.
   a What is the capacitance of the plates?
   b If the space between the plates is filled with titanium dioxide ($K = 100$), what is the capacitance of the plates?

6. A 10 μf capacitor is charged by applying a potential of 20 volts between the plates.
   a How much charge is there on each plate?
   b How much electrical work is done by the external 20 volt source in charging the capacitor?

1 a
  b
  c

2 a
  b

3

4

5 a
  b

6 a
  b

BASIC FACTS

The **electric potential** V, commonly referred to as **voltage**, at a point due to a charge q is,

$$V = \frac{1}{4\pi\epsilon_0 r}\frac{q}{r}, \qquad (15\text{-}1)$$

where r is the distance from the point to q and $\epsilon_0 = 8.854 \times 10^{-12}$ coul^2/nt m^2. If the charge is dispersed throughout a volume, then,

$$V = \frac{1}{4\pi\epsilon_0}\int\frac{dq}{r}. \qquad (15\text{-}1a)$$

The **electric field** E is related to the electric potential V by,

$$\mathbf{E} = -\frac{dV}{dr} \qquad (15\text{-}2)$$

The components of E in the x, y, and z directions are given by,

$$\mathbf{E}_x = -\frac{dV}{dx}, \quad (y, z = \text{const}) \qquad (15\text{-}2a)$$

$$\mathbf{E}_y = -\frac{dV}{dy}, \quad (x, z = \text{const})$$

$$\mathbf{E}_z = -\frac{dV}{dz}. \quad (x, y = \text{const})$$

This means that in a plot of V versus x, y, or z that $\mathbf{E}_x$, $\mathbf{E}_y$, or $\mathbf{E}_z$ at any value of x, y, or z, respectively, is the slope of the curve at that point.

If E is constant and makes a constant angle over the entire path shown in Fig. 15-2, then,

$$V_b - V_a = -\mathbf{E}\cdot\mathbf{s}, \qquad (15\text{-}3)$$

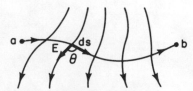

The electric field is
constant along path ab.

Fig. 15-2

(Continued on page 60)

ADDITIONAL INFORMATION

The electric force is very similar in form to the gravitational force (Eq. 8-1). Note that for point masses and point charges, F is proportional to $1/r^2$. The electrostatic energy of a charge q_1 a distance r from q_2 is,

$$W_{12} = \frac{1}{4\pi\epsilon_0}\frac{q_1 q_2}{r} \qquad (15\text{-}5)$$

This is similar in form to gravitational potential energy (Eq. 8-3).

Example 1. In Fig. 15-3, E is constant and three paths 1, 2, and 3 are shown. Since $\mathbf{E}\cdot\mathbf{s}$ varies a great

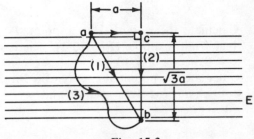

Fig. 15-3

deal along path 3, calculation of $V_b - V_a$ using this path would be extremely difficult. Along path 2, during the first part of the path, $V_c - V_a = -\mathbf{E}\cdot\mathbf{s} = -Es\cos 0° = -Es = -Ea$, where a is the distance from a to c; along the second part of the path, $V_b - V_c = -\mathbf{E}\cdot\mathbf{s} = -Es\cos 90° = 0$; thus, $V_b - V_a = (V_c - V_a) + (V_b - V_c) = -Ea$. Along path 1, $\mathbf{E}\cdot\mathbf{s} = Es\cos 60° = 0.5\,Es$; thus, $V_b - V_a = (-0.5E)(2a) = -Ea$, since the path length is $2a$.

Example 2. Calculate the potential at a distance z along the axis of a charged ring of radius a. (See Fig. 14-1.) dq and r are defined in example 2, chapter 14. It is not necessary to integrate because every part of the charge distribution is the same distance from the point in question; thus, $V = (1/4\pi\epsilon_0)Q/r = (1/4\pi\epsilon_0)$ $Q/\sqrt{z^2 + a^2}$.

A **capacitor** (or condenser) consists of two conductors separated by a **dielectric substance**, which acts as an *insulator*. **Capacitance**, which is the measure of a capacitor's ability to store a charge, is given by,

$$C = K\epsilon_0\frac{A}{d}, \qquad (15\text{-}6)$$

(Continued on page 60)

EXPLANATIONS

1. a $E = \dfrac{\sigma}{\epsilon_0} = \dfrac{5 \times 10^{-4} \text{ coul}/m^2}{8.9 \times 10^{-12} \text{ farad/m}} = 5.6 \times 10^7$ nt/coul
   (chapter 14), hence $F = qE = (-10^{-5} \text{ coul})(5.6 \times 10^7 \text{ nt/coul}) = -560$ nt at both positions.

   b $V = -E \cdot s = -(5.6 \times 10^7 \text{ nt/coul})(0.5 \text{ m}) = -2.8 \times 10^7$ volts.

   c $W = qV = (-10^{-5} \text{ coul})(-2.8 \times 10^7 \text{ volts}) = 280$ joule.

2. a $V = V_1 - V_2 = \dfrac{q}{4\pi\epsilon_0}\left(\dfrac{1}{r_1} - \dfrac{1}{r_2}\right) = \dfrac{q(r_2 - r_1)}{4\pi\epsilon_0 r_1 r_2}$,
   but $r_2 - r_1 = a\cos\theta$ and $r_1 r_2 \cong r^2$; thus,
   $$V = \dfrac{qa\cos\theta}{4\pi\epsilon_0 r^2}$$
   $$= \dfrac{(9 \times 10^9 \text{ nt m}^2/\text{coul}^2)(0.5 \text{ coul})(10^{-6} \text{ m})\cos\theta}{1 \text{ m}^2}$$
   $$= 4.5 \times 10^3 \cos\theta \text{ volts.}$$

   b $W = qV = (1 \text{ coul})(4.5 \times 10^3 \cos 60° \text{ volts}) = 2.25 \times 10^3$ joule. $(\cos 60° = \tfrac{1}{2}.)$

3. From Gauss's law (chapter 14), $E = 0$ inside and $E = \dfrac{q}{4\pi\epsilon_0 r^2}$ outside; thus, $V = -\displaystyle\int_\infty^R E\,dr = -\dfrac{q}{4\pi\epsilon_0}\int_\infty^R \dfrac{dr}{r^2} = \dfrac{q}{4\pi\epsilon_0 R}$ outside, because $V_\infty = 0$. Since $E = 0$ inside, $V = $ const inside, thus $V = \dfrac{q}{4\pi\epsilon_0 R}$ inside.

4. As in problem 3, $V = \dfrac{q}{4\pi\epsilon_0 r}$ outside; inside, however, $E = \dfrac{qr}{4\pi\epsilon_0 R^3}$. Therefore, $V = -\displaystyle\int_R^r E\,dr = -\int_R^r \dfrac{qr\,dr}{4\pi\epsilon_0 R^3} = -\dfrac{qr^2}{8\pi\epsilon_0 R^3}\Big|_R^r = \dfrac{q}{8\pi\epsilon_0}\left(-\dfrac{r}{R^3} + \dfrac{1}{R}\right).$

5. a $C = \dfrac{\epsilon_0 A}{d} = \dfrac{(8.9 \times 10^{-12} \text{ farad/m})(10 \text{ m}^2)}{0.01 \text{ m}} = 8.9 \times 10^{-7}$ farad $= 0.89$ μf.

   b $C = K\epsilon_0 A/d = (100)(8.9 \times 10^{-9} \text{ farad/m}) = 0.89$ μf.

6. a $q = CV = (10^{-5} \text{ farad})(20 \text{ volts}) = 2 \times 10^{-4}$ coul.

   b $W = \tfrac{1}{2}qV = \tfrac{1}{2}(2 \times 10^{-4} \text{ coul})(20 \text{ volts})$
      $= 2 \times 10^{-3}$ joule.

Answers

-560 nt	1 a
-2.8×10^7 volts	b
280 joule	c
$4.5 \times 10^3 \cos\theta$ volts	2 a
2.25×10^3 joule	b
$q/4\pi\epsilon_0 R$	3
$(q/8\pi\epsilon_0)(-r/R^3 + 1/R)$	4
0.89 μf	5 a
0.89 μf	b
2×10^{-4} coul	6 a
2×10^{-3} joule	b

where **s** is the length of the path, and V_a and V_b are the potentials at points a and b, respectively. Otherwise,

$$V_b - V_a = -\int_a^b \mathbf{E} \cdot d\mathbf{s}. \quad (15\text{-}3\text{a})$$

Since $V_b - V_a$ (Eq. 15-3 and 15-3a) does *not* depend on the path, the simplest path available can be selected.

If the calculation of the force on a charge q at a point is required, it is often easier to calculate **E** at the point and find **F** from $\mathbf{F} = q\mathbf{E}$. Similarly, for the potential of a charge q at a point, it is simpler to calculate V at the point using Eq. 15-9.

Only differences in potential can be measured; thus, V_b in Eq. 15-3 is measured relative to the potential at V_a. A point for which $V = 0$, called the **reference level**, must be assigned for every problem. Two such choices are common: 1) If the potential of a point infinitely far away is chosen to be zero, the potential of that point is measured relative to the potential *at infinity*. This choice was made in Eq. 15-1 and 15-1a. 2) If a nearby body is chosen to have zero potential, this body is said to be a **ground**. Any point connected to the ground by a conductor is said to be *grounded*. For example, if an automobile chassis is chosen for the ground, the positive pole is +12 volts relative to the chassis. This choice is commonly made in solving circuit problems (chapter 16).

The charge q on a body is proportional to the potential V of the body relative to the reference level; thus,

$$q = CV, \quad (15\text{-}4)$$

where C is the proportionality constant called the **capacitance**.

The unit of electrical potential is the volt: 1 volt = 1 joule/coul. The unit of capacitance is the farad: 1 farad = 1 coul/volt. 10^{-6} farad = 1 microfarad (μf or μfd); 10^{-12} farad = 1 micromicrofarad ($\mu\mu$f or $\mu\mu$fd). ϵ_0 can also be expressed in terms of farads: $\epsilon_0 = 8.85 \times 10^{-12}$ farad/m = 8.85 $\mu\mu$f/m.

where K is the dielectric constant, A is the area of the conductors, and d is the distance between them. The value of K differs according to the substance used for the dielectric, which may be a vacuum ($K = 1$), air ($K = 1.0006 \cong 1$), mica ($K = 5.7$), or shellac ($K = 3.0$), and various other substances. If two charges q_1 and q_2 are immersed in a fluid dielectric, the force F between them is given by,

$$F = \frac{q_1 q_2}{4\pi K \epsilon_0 r^2}. \quad (15\text{-}7)$$

Notice that F is reduced by a factor of $1/K$. A common type capacitor is a parallel-plate capacitor which consists of two metal plates separated by a dielectric. The electric intensity between the charged plates is given by,

$$E = \frac{\sigma}{\epsilon_0}, \quad (15\text{-}8)$$

where $\sigma = q/A$ is the charge density.

The **electric potential energy** required to charge a capacitor is,

$$W = \frac{1}{2} qV = \frac{1}{2} CV^2 = \frac{1}{2} \frac{q^2}{C}, \quad (15\text{-}9)$$

where V is the electrical potential at the location of q. The electric energy w_E per unit volume is,

$$w_E = \frac{1}{2} \epsilon_0 E^2. \quad (15\text{-}10)$$

The total electric energy in a volume V is,

$$W_E = \frac{1}{2} \epsilon_0 E^2 V, \quad (15\text{-}10\text{a})$$

if E is uniform in V; otherwise,

$$W_E = \int w_E dV, \quad (15\text{-}10\text{b})$$

where the integral is taken over the entire volume V.

Example 3.   Calculate the electric field energy for a capacitor if the plates are of area A and are separated by a distance d of vacuum. $W_E = \frac{1}{2} \epsilon_0 E^2 V = \frac{1}{2} \epsilon_0 E^2 Ad$; from Gauss's law for the electric field flux, Eq. 14-4, $E = q/\epsilon_0 A$ (using A instead of S), then, by substitution, $W_E = \frac{1}{2} \epsilon_0 (q/\epsilon_0 A)^2 Ad = \frac{1}{2} q^2 d/\epsilon_0 A = \frac{1}{2} q^2/C$, which agrees with Eq. 15-9.

Example 4.   Calculate the electric energy W_E in the region between two spherical shells of radii R_1 and R_2 and of charges q and $-q$. Since $E = (1/4\pi\epsilon_0)(q/R^2)$ for $R < R_1$ and $E = 0$ for $R < R_1$ and $R > R_2$, $w_E = \frac{1}{2} \epsilon_0 E^2 = (1/32\pi^2\epsilon_0)(q/R^4)$; also, since $w_E \neq$ const, Eq. 15-10b must be used, where $dV = -(q/4\pi\epsilon_0 R^2)$ from Eq. 15-1.

16

D. C. CIRCUITS

SELF-TEST

1. a Find the current supplied by the battery to the circuit in Fig. 16-1.
   b How much power does the battery supply this circuit?

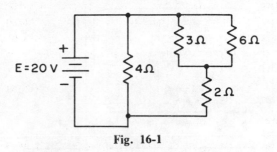

Fig. 16-1

2. A certain 50Ω (ohm) resistor carries 10 amp. How much power does the resistor dissipate?

3. A piece of wire 2000 m in length has a cross-sectional area of 0.01 m² and a resistance of 6Ω. Find the resistivity of the material.

4. If you were given four resistors of equal value, how would you connect them to obtain:
   a Maximum resistance?
   b Minimum resistance?

5. If you were given four capacitors of equal value, how would you connect them to obtain:
   a Maximum capacitance?
   b Minimum capacitance?

1 a
  b

2

3

4 a
  b

5 a
  b

61

BASIC FACTS

An **electric current** I is a flow of charges. Examples of currents are negative electrons flowing in a metal rod, positive and negative ions moving in a liquid, and a beam of electrons traveling in a gas or vacuum. The mks unit for current is the ampere (amp): 1 amp = 1 coul/sec; 1 milliampere (ma) = 10^{-3} amp; 1 microampere (μa) = 10^{-6} amp. A current can exist in a conductor only when an electric force is applied to the charges in the conductor. The agency which produces the electrical force necessary to move the charges is called a *seat* or *source* of **electromotive force**, abbreviated **emf**. The units of emf are volts (V), not nt.

A source of emf is a device which converts nonelectrical energy into electrical energy, such as a battery, which changes chemical energy into electrical energy, or a generator, which changes mechanical energy into electrical energy. The power P supplied by a source of emf is,

$$P = VI \qquad (16\text{-}1)$$

The direction of current is the direction in which positive charges appear to be moving. In a metal wire, only the electrons (negative charges) move, thus the direction of current flow is opposite to that of electron flow.

A resistor is a device which resists the flow of charges. If a current passes through a resistor, a potential difference is created between the ends of, hence *across*, the resistor. Resistance R is measured in ohms: 1 ohm (Ω) = 1 volt/amp, 1 megohm (M or meg) = 10^6 ohms, 1 kilohm (K) = 10^3 ohms.

The potential, or voltage, drop across a resistance is ΔV. **Ohm's law** states that,

$$\Delta V = IR. \qquad (16\text{-}2)$$

(Continued on page 64)

ADDITIONAL INFORMATION

Many circuit problems involve finding the current in a branch of a circuit or the voltage drop across a circuit element.

Example 1. Find the currents through each path and the voltage drops across each resistor in the circuit shown in Fig. 16-3. First, determine the volt-

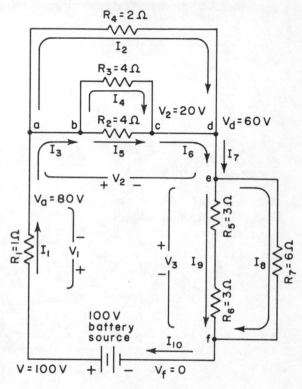

Fig. 16-3

age drops across the series resistances consisting of resistances R_1, $R_{ad} = (R_2, R_3, R_4)$, and $R_{ef} = (R_5, R_6, R_7)$: $R_1 = 1\ \Omega$ (ohm), $R_{ad} = 1/(1/R_2 + 1/R_3 + 1/R_4) = 1/(1/4\Omega + 1/4\Omega + 1/2\Omega) = 1\Omega$, $R_{ef} = 1/\left(\dfrac{1}{R_5 + R_6} + \dfrac{1}{R_7}\right) = 1/(1/6\Omega + 1/6\Omega) = 3\Omega$. Thus, $R_{\text{tot}} = 5\Omega$. Using Ohm's law $I = V/R$, the total current delivered to the circuit by the battery is $I_{\text{tot}} = 100\text{V}/5\Omega = 20$ amp, therefore, the voltage drop across R_1 is $I_{\text{tot}}R_1 = (20\text{ amp})(1\Omega) = 20\text{V}$, the voltage drop across R_{ad} is $I_{\text{tot}}R_{ad} = (20\text{ amp})(1\Omega) = 20\text{V}$, and the voltage drop across R_{ef} is $I_{\text{tot}}R_{ef} = (20\text{ amp})(3\Omega) = 60\text{V}$. The voltage at point a is, therefore, $V_{\text{tot}} - V_{R_1} = 100\text{V} - 20\text{V} = 80\text{V}$; similarly, the voltage at point d is $V_{\text{tot}} - V_{R_1} - V_{R_{ad}} = 100\text{V} - 20\text{V} - 20\text{V} = 60\text{V}$; hence, the voltage drop across

(Continued on page 64)

EXPLANATIONS

1. a The circuit in Fig. 16-1a can be converted to the circuit in Fig. 16-1b by combining the resistors in the right branch. The two circuits still have the same resistance as far as the battery is concerned. The 3Ω and 6Ω resistors are in *parallel*; thus their total resistance is 2Ω. This resistance is in series with 2Ω, which adds up to 4Ω. This 4Ω is in parallel with 4Ω so that the total resistance $R_{tot} = 2\Omega$. The current is
$$I = V/R = 20V/2\Omega = 10 \text{ amp.}$$

   b $P = IV = (10 \text{ amp})(20V) = 200 \text{ watts.}$

2. $P = I^2R = (10 \text{ amp})^2(50\Omega) = 5000 \text{ watts.}$

3. $\rho = \dfrac{RA}{L} = \dfrac{(6\Omega)(0.01 \text{ m}^2)}{(2000 \text{ m})} = \dfrac{(6\Omega \times 10^{-2} \text{ m}^2)}{(2 \times 10^3 \text{ m})}$
    $= 3 \times 10^{-5} \, \Omega\text{m.}$

4. a In series.
    b In parallel.

5. a In parallel.
    b In series.

Answers

10 amp	1 a
200 watts	b
5000 watts	2
$3 \times 10^{-5} \, \Omega \text{ m}$	3
In series	4 a
In parallel	b
In parallel	5 a
In series	b

Common Circuit Symbols

resistor —⌇⌇⌇—
variable resistors:
    rheostat —⌇⌇⌇—
    potentiometer —⌇⌇⌇—
capacitor ⊣⊢
battery ⊣ı⊦
inductor (See chapter 18.) ⌒⌒⌒⌒⌒
ground point ⊥

Capacitors in series

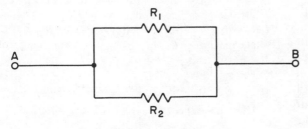

Resistors in parallel

Fig. 16-2

A resistor carrying a current heats up because it converts electrical energy into heat. The amount of energy converted per unit time is the electric power P dissipated; thus,

$$P = I\Delta V = I^2 R = \frac{\Delta V^2}{R}. \quad (16\text{-}3)$$

A resistor of length L and area A with resistance R has **resistivity** ρ, where,

$$\rho = \frac{RA}{L} \quad (16\text{-}4)$$

Kirchhoff's laws state that: 1) The algebraic sum of the currents entering any junction point at which two or more wires are connected is zero, and 2) The algebraic sum of the changes in potential around any closed path is zero. A potential change may be positive or negative.

Two elements are *in series* if they have only one common terminal and no other circuit elements are connected to the common terminal. Two elements are *in parallel* if they have two common terminals (Fig. 16-2). The total resistance R_{tot} of n resistors connected *in series* is,

$$R_{\text{tot}} = R_1 + R_2 + \cdots + R_n. \quad (16\text{-}5)$$

The total resistance of n resistors connected *in parallel* is,

$$R_{\text{tot}} = \frac{1}{\dfrac{1}{R_1} + \dfrac{1}{R_2} + \cdots + \dfrac{1}{R_n}}. \quad (16\text{-}6)$$

The unit of electric power is the watt in the mks system: 1 watt (w) = 1 joule/sec = 1 (joule/coul) (coul/sec) = 1 volt amp; 1 megawatt (Mw) = 10^6 watts, 1 kilowatt (kw) = 10^3 watts, 1 milliwatt (mw) = 10^{-3} watt, 1 watt sec = 1 joule, and 1 kilowatt hour (kwh) = 3.6×10^6 joule. Note that the *watt second* and the *kilowatt hour* are units of *energy*.

R_5 and R_6 is 30V + 30V = 60V, and the voltage drop at point f is $V_{\text{tot}} - V_{R_1} - V_{R_{ad}} - V_{R_{ef}} = 100\text{V} - 20\text{V} - 20\text{V} - 60\text{V} = 0\text{V}$. The current through each path can now be computed: By *Kirchhoff's first law*, a total of 20 amp will flow through resistances R_1, R_{ad}, and R_{ef}. Thus, 20 amp flows through R_1. Recalling that there is a 20V drop across R_{ad} and applying *Ohm's law*, $I_2 = V_2/R_4 = 20\text{V}/2\Omega = 10$ amp, hence $I_3 = I_1 - I_2 = 20$ amp $- 10$ amp $= 10$ amp, $I_4 = I_3 - I_5 = I_3 - V_{R_2}/R_2 = 10$ amp $- 5$ amp $= 5$ amp, hence $I_5 = 5$ amp. Thus, there are 10 amp entering point b and 10 amp leaving point b. Therefore, $I_6 = I_4 + I_5 = 10$ amp and $I_7 = I_2 + I_6 = 20$ amp. Since 20 amp enters point e and leaves point f and $R_5 + R_6 = R_7 = 6\Omega$, 10 amp flows through both I_8 and I_9, and finally $I_{10} = 20$ amp flows to source.

The Wheatstone bridge is used to measure the values of unknown resistances. In Fig. 16-4, G is a

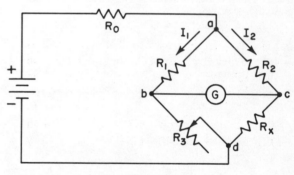

Fig. 16-4

meter movement called a galvanometer that measures current flow, R_x is the unknown resistor, and R_3 is a resistor which can be adjusted until no current flows through the galvanometer. When current flows through the bridge, the meter registers zero. The bridge is then *balanced* and a relationship between R_x and R_3 can be found. Using Ohm's law, $I_1(R_1 + R_3) = V_{ad} = I_2(R_2 + R_x)$. Also, since there is no current through the meter, the voltage at points b and c must be equal. Thus, $V_{ab} = V_{ac}$, so that $I_1 R_1 = I_2 R_2$. Therefore, $I_1 R_3 + I_2 R_x$, and $R_x = R_3(I_1/I_2) = R_3(R_2/R_1)$. The total capacitance C_{tot} of n capacitors *in series* is,

$$C_{\text{tot}} = \frac{1}{\dfrac{1}{C_1} + \dfrac{1}{C_2} + \cdots + \dfrac{1}{C_n}}. \quad (16\text{-}7)$$

The capacitance of n capacitors *in parallel* is,

$$C_{\text{tot}} = C_1 + C_2 + \cdots + C_n \quad (16\text{-}8)$$

The total inductance L_{tot} of a set of n inductors (chapter 18) is calculated the same as resistance.

17

MAGNETIC FIELDS

1. A solenoid 0.5 m long has 10^3 turns of wire and carries 1 amp of electric current. What is the magnetic field inside the solenoid?

2. A bar magnet has north and south poles of 10^{-4} amp m; its length is 5 cm. The magnet is placed perpendicularly to a B field of 10 weber/m^2.
   a Find the magnetic dipole moment for the magnet.
   b What is the torque on the magnet?
   c What is the potential energy of the magnet?

3. The magnet in problem 2 is placed antiparallel to the field. (The magnetic dipole is lined against the field.) Calculate the potential energy and torque of the magnet.

4. The magnet in problem 2 is now lined with the field. Calculate the potential energy and torque of the magnet.

5. A very wide sheet of copper carries a current of i amp/cm width (Fig. 17-1). Calculate the magnetic field near the sheet.

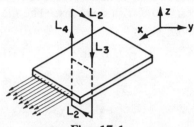

Fig. 17-1.

6. Calculate the magnetic field with respect to the distance r from the axis of a solid wire of radius R which carries a current that is uniformly distributed throughout the cross section of the wire.

7. Calculate the magnetic field with respect to the distance r from the axis of a cylindrical shell of radius R carrying a current parallel to the axis.

1 _____

2 a _____
  b _____
  c _____

3 _____

4 _____

5 _____

6 _____

7 _____

BASIC FACTS

The force on a wire of length L carrying a current I in a **magnetic field** B (Fig. 17-2) is,

$$\mathbf{F} = I\mathbf{L} \times \mathbf{B}, \qquad (17\text{-}1)$$

where $\mathbf{L}$ is directed along I and θ is the angle between $\mathbf{L}$ and $\mathbf{B}$, $\mathbf{L}$ is in meters, $\mathbf{B}$ is in webers/m^2 (mks units), I is in amperes, and $\mathbf{F}$ is in newtons. The magnitude of $\mathbf{F}$ is,

$$F = ILB \sin \theta. \qquad (17\text{-}1a)$$

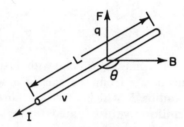

Fig. 17-2.

The force on a single charge q moving with velocity v in a magnetic field B is,

$$\mathbf{F} = q\mathbf{v} \times \mathbf{B}, \qquad (17\text{-}2)$$

where θ is the angle between $\mathbf{v}$ and $\mathbf{B}$. The magnitude of $\mathbf{F}$ is,

$$F = qvB \sin \theta. \qquad (17\text{-}2a)$$

Also,

$$\mathbf{F} = m\mathbf{B}, \qquad (17\text{-}3)$$

and the force exerted on a magnetic pole m_1 by a magnetic pole m_2 is,

$$\mathbf{F} = \frac{\mu_0}{4\pi} \frac{m_1 m_2}{r^2}. \qquad (17\text{-}4)$$

Poles of like sign repel one another and those of opposite sign attract. **North poles** are *positive* and **south poles** are *negative*, by convention. A positive pole cannot exist without a nearby negative pole of equal magnitude. A north pole of magnitude $+m$ and a south pole of

(Continued on page 68)

ADDITIONAL INFORMATION

Magnetic fields are produced by currents and magnetic poles. The magnetic field dB produced by a current I in a wire of length dL is,

$$d\mathbf{B} = \frac{\mu_0 I}{4\pi} \frac{d\mathbf{L} \times \mathbf{e}}{r^2}, \qquad (17\text{-}9)$$

where $\mu_0 = 4\pi \times 10^{-7}$ weber/amp m, $\mathbf{e}$ is the unit vector directed from $d\mathbf{L}$ to the point at which $\mathbf{B}$ is being calculated, and θ is the angle between $\mathbf{e}$ and $d\mathbf{L}$ (Fig 17-3). The magnitude of $d\mathbf{B}$ is,

$$dB = \frac{\mu_0 I}{4\pi} \frac{dL \sin \theta}{r^2}. \qquad (17\text{-}9a)$$

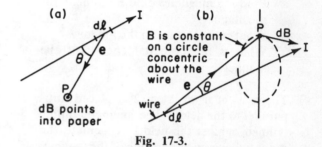

Fig. 17-3.

The total magnetic field B at a point is,

$$B = \frac{\mu_0}{4\pi} \int \frac{I}{r^2} \sin \theta \, dL \qquad (17\text{-}10)$$

The magnetic field resulting from a magnetic pole m is,

$$B = \frac{\mu_0}{4\pi} \frac{m}{r^2}, \qquad (17\text{-}10a)$$

where m is in amp m and r is the distance from m to the point at which B is also being calculated.

A solenoid is a cylindrical coil of wire; the number of turns per unit length is called n. The magnetic field B inside a long solenoid is essentially constant; thus,

$$B = \mu_0 n I. \qquad (17\text{-}10b)$$

An electric field exerts a force F on any charge whether or not it is moving. Recalling that $\mathbf{F} = q\mathbf{E}$, note that $\mathbf{F}$ is in the same direction as $\mathbf{E}$. A magnetic field exerts a force only on currents (moving charges) and magnetic poles. A single charge q (coul) moving with velocity $\mathbf{v}$ (m/sec) is an important special case of a current because $I\mathbf{L} = q\mathbf{v}$. The units balance, since (coul/sec)(m) = (coul)(m/sec). The velocity vector

(Continued on page 68)

EXPLANATIONS

1. $B = \mu_0 \, nI$

$$= (4\pi \times 10^{-7} \text{ weber/amp m}) \, \frac{10^3 \text{ turns}}{0.5 \text{ m}} \text{ (1 amp)}$$

$$= 8\pi \times 10^{-4} \text{ weber/m}^2.$$

2. a $\quad p_m = md = (10^{-4} \text{ amp m})(0.05 \text{ m}) = 5 \times 10^{-6} \text{ amp m}^2$.
   b $\quad \tau = p_m B \sin \theta = (5 \times 10^{-6} \text{ amp m}^2)(10 \text{ weber/m}^2)$ $(\sin 90°) = 5 \times 10^{-5} \text{ amp weber} = 5 \times 10^{-5} \text{ nt m}$.
   c $\quad U = p_m B \cos \theta = 0$, because $\cos 90° = 0$.

3. $U = -p_m B \cos \theta = -(5 \times 10^{-6} \text{ amp m}^2)(10 \text{ weber/m}^2)$ $(\cos 180°) = 5 \times 10^{-5}$ joule, and $\tau = p_m B \sin 180° = 0$.

4. $U = -p_m B \cos 0° = -5 \times 10^{-5}$ joule; $\tau = p_m B \sin 0° = 0$.

5. In Fig. 17-1 the amperian path pierces the sheet; B should be parallel to L_1 and L_2, and perpendicular to L_3 and L_4. The current passing through the path is iL_1. Therefore, by Ampere's law, $BL_1 + BL_2 = 2BL_1 = \mu_0 iL_1$ and $B = \mu_0 i/2$. (B = const near the sheet.)

6. **B** is directed along the circular path around the wire (Fig. 17-5). If $r > R$, the radius of the wire, then the current through

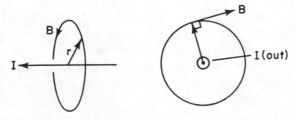

Fig. 17-5.

the surface is I, and Ampere's law states that $B(2\pi r) = \mu_0 I$ and $B = \mu_0 I/2\pi r$; however, if $r < R$, I_r through the path is proportional to the area of the path; hence, $I_r = I\pi r^2/\pi R^2 = Ir^2/R^2$, so that $B = \mu_0 Ir/2\pi R^2$.

7. The magnetic field B outside the shell of radius R ($r > R$) is the same as that in problem 6; however, inside the shell ($r < R$) I through L is zero, therefore, $B = 0$.

Answers

$8\pi \times 10^{-4}$ weber/m^2	1
5×10^{-6} amp m^2	2 a
5×10^{-5} nt m	b
0	c
5×10^{-5} joule, 0	3
5×10^{-5} joule, 0	4
$\mu_0 \, i/2$	5
$\mu_0 I/2\pi r$ $(r > R)$, $\mu_0 I_r/2\pi R^2$ $(r < R)$	6
$\mu_0 I/2\pi r$ $(r > R)$, 0 $(r < R)$	7

magnitude $-m$ separated by a distance d constitute a **magnetic dipole** p_m, where,

$$p_m = md. \qquad (17\text{-}5)$$

The vector p_m is directed from the south pole to the north pole.

A **current loop** is a magnetic dipole in which p_m is directed along the *normal* to the area S of the loop that carries the current I so that,

$$p_m = IS. \qquad (17\text{-}5a)$$

The **torque** τ on a magnetic dipole in a magnetic field B is,

$$\tau = p_m \times \mathbf{B}. \qquad (17\text{-}6)$$

The magnitude of τ is,

$$\tau = p_m B \sin \theta. \qquad (17\text{-}6a)$$

The **potential energy** U of a dipole in a magnetic field is,

$$U = -p_m \cdot \mathbf{B} = -p_m B \cos \theta. \qquad (17\text{-}7)$$

Ampere's law states that if a magnetic field B is constant on a closed path of length L, then,

$$\mathbf{B} \cdot \mathbf{L} = BL \cos \theta = \mu_0 I, \qquad (17\text{-}8)$$

where θ is the angle between $\mathbf{B}$ and $\mathbf{L}$, and I is the current that "pierces" L. If $B \neq$ const on L, then,

$$\oint \mathbf{B} \cdot d\mathbf{L} = \mu_0 I. \qquad (17\text{-}8a)$$

The circle on the integral sign indicates that the path is closed.

The unit of flux is the weber (mks); the unit of magnetic induction is the weber per square meter: 1 weber/m^2 = 1 nt sec/coul m = 1 nt/amp m; also, 1 weber/m^2 = 10^4 gauss (cgs).

v has two components: $v_{\parallel}$ is the component *parallel* to $\mathbf{B}$ and $v_{\perp}$ is the component *perpendicular* to $\mathbf{B}$, where $\mathbf{B}$ exerts a force only on $v_{\perp}$. $\mathbf{B}$ is often shown pointing into the plane by $\oplus$ and out of the plane by $\odot$.

In Fig. 17-4, the force on a positive charge q moving in the plane of the paper is perpendicular to the mag-

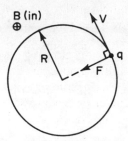

Fig. 17-4.

netic field, which is away from ($\oplus$) the reader. Since the path of q is a circle of radius R, the acceleration must be, $a_c = v^2/R$. From Newton's second law $F = ma$, where m is the mass of the charge; hence,

$$R = \frac{mv}{qB}; \qquad (17\text{-}11)$$

also,

$$v = \frac{2\pi R}{T}, \qquad (17\text{-}11a)$$

hence,

$$T = \frac{2\pi R}{v} = \frac{2\pi m}{qB} \qquad (17\text{-}11b)$$

Note that the period T is independent of the velocity v and the radius R. Thus, no matter how fast a charge moves in a uniform magnetic field, the period will not change.

Example 1.   Calculate the force of attraction between a -1 amp m south pole and a 1 amp m north pole separated by 1 m and compare this force to the force between a $+1$ coul charge and a -1 coul charge separated by one meter.

$$F_{mag} = \frac{\mu_0}{4\pi} \frac{m_1 m_2}{r^2}$$

$$= \frac{(10^{-7}\text{weber/amp m})(1 \text{ amp m})(-1 \text{ amp m})}{(1 \text{ m})^2}$$

$$= -10^{-7} \text{ nt};$$

$$F_{el} = \frac{1}{4\pi\epsilon_0} \frac{q_1 q_2}{r^2}$$

$$= (9 \times 10^9 \text{ nt m}^2/\text{coul}^2) \frac{(1 \text{ coul})(-1 \text{ coul})}{(1\text{m})^2}$$

$$= -9 \times 10^9 \text{ nt}.$$

1. A coil with 500 turns and an area of 0.1 m^2 in a magnetic field is turning at 2 Hz (Hertz = revolutions per sec)(Fig. 18-1). The magnetic field is 0.1 weber/m^2.
   a   Express the flux through the coil with respect to the angle θ in Fig. 18-2.
   b   Express θ and the magnetic flux as time-dependent functions.
   c   Find the emf generated in the coil.

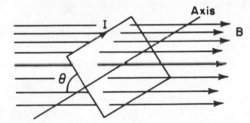

Fig. 18-1

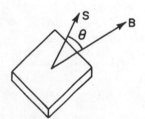

Fig. 18-2

2. In Fig. 18-3, wire A is pulled along U-shaped wire B, the ends of which are 10 cm apart. Wire A is pulled at a velocity of 20 cm/sec, the resistance of wire B is 1 Ω, and the magnetic flux is everywhere 1 weber/m^2 and is normal to the plane S containing the wires.
   a   Using Faraday's law, calculate the emf induced around the circuit.
   b   Calculate the magnitude and direction of the current induced in the circuit formed by wires A and B.
   c   Calculate the force (in newtons) on 1 coul of electrons (negative charges) in wire A moving with a velocity of 20 cm/sec.

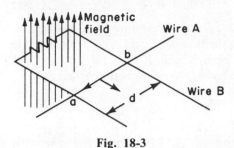

Fig. 18-3

   d   Electrons are moved by a force and produce a current. Calculate the difference in potential across the ends of wire A and the current flow through the circuit.
   e   Are the results obtained in *d* the same as those obtained in *a* and *b*?
   f   Assuming that wire A slides frictionlessly on wire B, calculate the force necessary to keep wire A moving at 20 cm/sec.

3. A very long cylindrical solenoid (Fig. 18-4) with 5000 turns per meter is the primary winding of a transformer with area 0.01 m^2; the secondary winding has an area of 0.02 m^2 and consists of 25 turns wrapped around the primary.
   a   Calculate the mutual inductance.
   b   If the primary current is 2 amp, what is the flux linkage in the secondary circuit?
   c   If the primary current changes from 2 amp to 4 amp in 10 sec, what is the average emf induced in the secondary?

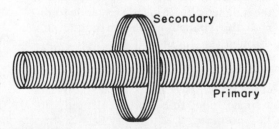

Fig. 18-4

1	a
	b
	c
2	a
	b
	c
	d
	e
	f
3	a
	b
	c

BASIC FACTS

The **magnetic flux** Φ through a surface is defined in a similar way to the electric flux described in chapter 14; thus,

$$\Phi = \mathbf{B} \cdot \mathbf{S} = BS \cos \theta, \quad (18\text{-}1)$$

where $\mathbf{S}$ is a vector normal to the surface, $\mathbf{B}$ is the magnetic field, and θ is the angle between $\mathbf{B}$ and $\mathbf{S}$ (Fig. 18-1). If $\mathbf{B}$, $\mathbf{S}$ and/or θ changes over the surface,

$$\Phi = \int_s \mathbf{B} \cdot d\mathbf{S}. \quad (18\text{-}1a)$$

If the flux Φ through S changes with time, an emf (chapter 16) is induced around the border of the surface.

Faraday's law states that the **induced emf** $\mathcal{E}$ equals the *negative* of the rate of change with respect to time of the magnetic flux within a circuit. Faraday's law is given by,

$$\mathcal{E} = \mathbf{E} \cdot \mathbf{l} = -\frac{\Delta \Phi}{\Delta t} = -\frac{\Delta(\mathbf{B} \cdot \mathbf{S})}{\Delta t}, \quad (18\text{-}2)$$

where $\mathbf{E}$ is the electric field and $\mathbf{l}$ is the distance around the surface. Also,

$$\mathcal{E} = \oint \mathbf{E} \cdot d\mathbf{l} = -\frac{d\Phi}{dt} = -\frac{d}{dt} \oint_s \mathbf{B} \cdot d\mathbf{S},$$
$$(18\text{-}2a)$$

which is the line integral around a closed path.

The emf develops an **induced current** $I = \mathcal{E}/R$ in a closed circuit as required by Ohm's law (chapter 16). The induced current also produces its own magnetic field and its own flux through the circuit.

Lenz's law states that the direction of the induced current is always such that the magnetic flux it produces tends to *oppose* the change in the magnetic flux through the circuit. If the circuit consists of a solenoid (coil of wire) with N turns, then the magnetic flux through the entire circuit is N times the flux through a single loop; thus, for a solenoid, Eq. 18-2 becomes, $\mathcal{E} = -N\Delta\Phi/\Delta t$.

(Continued on page 72)

ADDITIONAL INFORMATION

A closed electric circuit has an area S and a perimeter (the path around a wire). If the magnetic flux changes with time, a voltage is induced around a closed circuit causing a current to flow. An emf is induced even if no circuit exists, but obviously no current (charge flow) is produced.

There are several ways to produce an emf in a circuit using Faraday's law:

1. An external magnetic field can be changed.
2. A magnet producing the external field can be moved or rotated, thus changing angle θ in Eq. 18-1.
3. A circuit can be moved from a region of high magnetic field to a region of low magnetic field.
4. The shape and surface area of a coil can also be changed.
5. The coil can be rotated also changing angle θ in Eq. 18-1.

Example 1. A circuit consisting of a circle of copper wire with a 4 cm radius and a 2000 Ω resistor is in a uniform magnetic field of 1 weber/m². The field is then turned off and it decays uniformly in 1 sec to zero. Calculate the current in the circuit. The magnetic flux $\Phi = \mathbf{B} \cdot \mathbf{S} = (1 \text{ weber/m}^2)(0.04 \text{ m})^2 = 1.6 \times 10^{-3}$ weber before the magnet is turned off; Φ is zero 1 sec after the magnet is turned off; thus,

$$\frac{\Delta \Phi}{\Delta t} = \frac{(\Phi_f - \Phi_i)}{1 \text{ sec}} = \frac{(0 - 1.6 \times 10^{-3}) \text{ weber}}{1 \text{ sec}}$$
$$= -1.6 \times 10^{-3} \text{ volt},$$

where the subscripts f and i stand for "final" and "initial," respectively. The current is $I = \mathcal{E}/R = (-1.6 \times 10^{-3} \text{ volt})/(2 \times 10^3 \text{ ohms}) = -8 \times 10^{-7}$ amp.

Example 2. A 10 cm × 10 cm solenoid, consisting of 5000 turns, is rotated 90° in a 2×10^4 gauss field in 4 sec. Calculate the average emf induced in the solenoid during the rotation. Initially, the flux $N\Phi = NBS \cos 90° = 0$; the maximum emf would be $N\Phi = NBS \cos 0° = NBS = (5000 \text{ turns}) (2 \text{ weber/m}^2) (0.01 \text{ m}^2) = 100$ webers. Thus,

$$\mathcal{E} = \frac{\Delta \Phi}{\Delta t} = \frac{100 \text{ webers}}{4 \text{ sec}} = -25 \text{ volts}.$$

The direction of the current is such that it opposes an increase in Φ, according to Lenz's law.

(Continued on page 72)

EXPLANATIONS

1. a $\Phi = BS\cos\theta = (0.1 \text{ weber}/m^2)(0.1 \text{ m}^2)\cos\theta = (0.01)$
   $\cos\theta$ weber.

   b $\theta = \omega t = 2\pi ft = 2\pi(2 \text{ Hz})t = 4\pi t$, since $\omega = 2\pi f$,
   and $\Phi = BS\cos\theta = (0.01)\cos 4\pi t$ weber.

   c $\mathcal{E} = -\dfrac{d\Phi}{dt} = -(0.01 \text{ volt sec})(-4\pi\sin 4\pi t) = (0.04\pi)$
   $\sin 4\pi t$ volt.

2. a $\mathcal{E} = -\dfrac{\Delta\Phi}{\Delta t} = -\dfrac{B\Delta S}{\Delta t} = Bvl = -(1 \text{ weber}/m^2)(0.2$
   m/sec)(0.1 m) $= -0.02$ volt.

   b $I = \dfrac{\mathcal{E}}{R} = -0.02 \text{ volt}/1\Omega = -0.02$ amp.

   c $F = qvB = (1 \text{ coul})(0.2 \text{ m/sec})(1 \text{ weber}/m^2)$
   $= 0.2$ nt. $(\theta = 90°)$ This is the magnitude of the force
   on 1 coul of electrons. Electrons flow from *a* to *b*, but
   current flow is from *b* to *a*.

   d $\mathcal{E} = -\dfrac{W}{q} = -\dfrac{Fl}{q} = -\dfrac{(0.2 \text{ nt})(0.1 \text{ m})}{1 \text{ coul}} = 0.02$ volt.
   $I = \mathcal{E}/R = 0.02$ amp.

   e The methods used in calculating *a, b, c,* and *d* are equivalent, therefore the results must agree.

   f $F = IlB = (0.02 \text{ amp})(0.1 \text{ m})(1 \text{ weber}/m^2) = 2 \times 10^{-3}$ nt. $(\theta = 90°)$.

3. a $B_1 = \mu_0 nI$ inside, and B_1 is essentially zero outside;
   thus, $B_1 = (4\pi \times 10^{-7} \text{ weber}/\text{amp m})(5 \times 10^3 \text{ turns/m})$
   $(I_1 \text{ amp}) = 2\pi I_1 \times 10^{-3} \text{ weber}/m^2$ inside. $\Phi_2 = \Phi_1 =$
   $B_1 S_1 = (2\pi I_1 \times 10^{-3} \text{ weber}/m^2)(1 \times 10^{-2} m^2) =$
   $2\pi I_1 \times 10^{-5}$ weber; therefore, $N_2\Phi_2 = 5\pi \times 10^{-4} I_1$
   weber $= MI_1$, and $M = 5\pi \times 10^{-4}$ weber/amp.

   b $N_2\Phi_2 = MI_1 = \pi \times 10^{-3}$ weber.

   c $\mathcal{E} = -\dfrac{N_2\Delta\Phi_2}{\Delta t} = -\dfrac{M\Delta I_1}{\Delta t}$

   $= -\dfrac{(5\pi \times 10^{-4} \text{ weber/amp})(2 \text{ amp})}{10 \text{ sec}}$

   $= -\pi \times 10^{-4}$ volt.

Answers

$(0.01)\cos\theta$ weber	1 a
$(0.01)\cos 4\pi t$ weber	b
$(0.04\pi)\sin 4\pi t$ volt	c
-0.02 volt	2 a
-0.02 amp	b
0.2 nt	c
-0.02 volt, 0.02 amp	d
yes	e
2×10^{-3} nt	f
$5\pi \times 10^{-4}$ weber/amp	3 a
$\pi \times 10^{-3}$ weber	b
$\pi \times 10^{-4}$ volt	c

If the magnetic field produced by one circuit, called the **primary circuit**, passes through another circuit, called the **secondary circuit,** then there is a **flux linkage** between the two circuits. The *primary* current I_1 produces a flux linkage $N_2\Phi_2$ (Subscripts 1 and 2 denote the primary and secondary circuits, respectively.), where N_2 is the number of turns if circuit is a solenoid. It has been found experimentally that $N_2\Phi_2$ is proportional to I_1. The **mutual inductance** M is given by the relationship.

$$N_2\Phi_2 = MI_1 \text{ and } N_1\Phi_1 = MI_2. \qquad (18\text{-}3)$$

Expressing Φ_1 as $B_1 S_1$ and using Faraday's law,

$$\frac{M\Delta I_1}{\Delta t} = \frac{N_2 B_2 S_2}{\Delta t} = -\mathcal{E}_2 \qquad (18\text{-}4)$$

and, $\dfrac{M\Delta I_2}{\Delta t} = \dfrac{N_1 B_1 S_1}{\Delta t} = -\mathcal{E}_1$

If I_1 changes, an emf is induced in the secondary circuit.

The flux $N\Phi$ in a circuit produced by a current I in the *same* circuit is,

$$N\Phi = LI, \qquad (18\text{-}5)$$

where L is the **self-inductance;** also,

$$\frac{L\Delta I}{\Delta t} = \frac{\Delta NBS}{\Delta t} = -\mathcal{E}. \qquad (18\text{-}6)$$

Here, $\mathcal{E}$ is the **self-induced emf.** When $I \neq$ const, the calculus must be used,

$$\mathcal{E} = -L\frac{dI}{dt}. \qquad (18\text{-}6a)$$

The unit of magnetic flux is the weber: 1 weber = 1 volt sec; the emf is in volts. The unit of inductance is the henry (mks): 1 henry = 1 volt/(amp/sec) = 1 volt/ (amp/sec) = 1 sec/amp = 1 weber/amp.

A pair of coils with a large mutual inductance serves as a transformer. An alternating current in the primary circuit induces a current in the coil in the secondary circuit. The voltages in the primary and secondary circuits are related to one another by $V_2/V_1 = N_2/N_1$. This equation is derived from Faraday's law, therefore, V_1 must be time dependent.

Example 3. A 110 volt AC source is connected to the primary of a transformer which has 100 turns. If the output voltage of the transformer is 220 volts, what is the number of turns on the secondary coil? The number of turns on the secondary $N_2 = N_1 V_2/V_1 = (100 \text{ turns})(220 \text{ volts})/(110 \text{ volts}) = 200$ turns.

The voltage in the primary can be "stepped up" or "stepped down," but the power output cannot be increased. In example 3, the voltage is stepped up by a factor of 2, but the current in the secondary must be less than the current in the primary by at least a factor of $\frac{1}{2}$. For an ideal transformer with 100% efficiency, the power output P_2 equals the input power P_1. Using the relationship $P = VI$, the power becomes $V_2 I_2 = V_1 I_1$, or $I_2/I_1 = V_2/V_1$. This result follows from the law of the conservation of energy. Mutual inductance and self-inductance are independent of currents and fluxes.

The energy W_B stored in an inductor is,

$$W_B = \frac{1}{2}LI^2 \qquad (18\text{-}7)$$

The energy w_B stored in a magnetic field per unit volume is,

$$w_B = \frac{1}{2}\frac{B^2}{\mu_0}. \qquad (18\text{-}8)$$

The magnetic energy in a volume V is,

$$W_B = Vw_B. \qquad (18\text{-}9)$$

If $B \neq$ const,

$$W_B = \int w_B dV. \qquad (18\text{-}9a)$$

19

ALTERNATING CURRENTS

1. The capacitor in Fig. 19-1 has two circular plates of areas 1.0 m² separated by 0.001 m of air; R is 2000 Ω.
   a What is the capacitance?
   b If the potential across the capacitor is initially 100V, what is the initial charge?
   c Find the charge on the capacitor.
   d Find the voltage across the capacitor.
   e Find the electric field and the electric flux across the capacitor.
   f Find the displacement current through the capacitor.
   g What is the magnetic field at the edge of the capacitor?

h What is the impedence, current, and phase angle, if a 2×10^4 Hz (Hertz), 300V source is connected between the resistor and capacitor?

2. The RLC circuit in Fig. 19-2 has a 20 Ω (ohm) resistor, a 20 μfd (microfarad) capacitor and a 5 henry inductor.
   a At what frequency will the current oscillate?
   b At what frequency will the amplitude of the oscillation be one half the original amplitude?

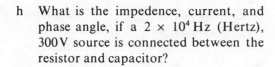

1	a
	b
	c
	d
	e
	f
	g
	h
2	a
	b

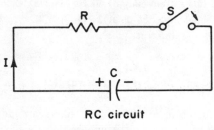

RC circuit

Fig. 19-1

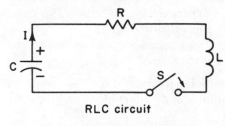

RLC circuit

Fig. 19-2

BASIC FACTS

The circuits considered in this chapter contain currents which change with time in electrical circuits containing resistance, capacitance, and inductance.

The **RC circuit** in Fig. 19-1 contains only resistance and capacitance (hence the name "RC") and has no source of emf. The capacitor C is charged at time $t = 0$ (the *initial condition*). At $t = 0$ the switch S is closed; the initial voltage drop across C is V_0 and the initial charge on C is q_0 (See Eq. 15-5.). The capacitor discharges, causing current to flow through resistor R and the charge q on C decreases *exponentially* with time (t),

$$q(t) = q_0 \operatorname{Exp}(-t/RC), \quad (19\text{-}1)$$

where $\operatorname{Exp} = e \cong 2.718$. Also, the current through R is $I(t) = (q/RC) \operatorname{Exp}(-t/RC)$. The charge and current both decay *exponentially* to zero. The electrical energy stored in the capacitor is dissipated (turned into heat) in the resistor at the rate of $P = I^2 R$.

The **RL circuit** in Fig. 19-3 contains resistance and inductance, and has no

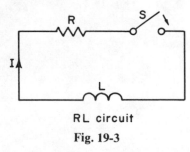

RL circuit

Fig. 19-3

source of emf. At $t = 0$ and $I = I_0$. The current varies *exponentially* with time,

$$I(t) = I_0 \operatorname{Exp}(-Rt/L). \quad (19\text{-}2)$$

The current also decays *exponentially* and dissipates the magnetic energy stored in the inductor L (Eq. 18-10) through the resistor at the rate of $P = I^2 R$.

The **LC circuit** in Fig. 19-4 contains both inductance and capacitance, but no

(Continued on page 76)

ADDITIONAL INFORMATION

For direct currents (DC) an ideal inductor has no resistance, but a capacitor has infinite resistance (*no current passes.*). A source of emf that oscillates sinusoidally (like a sine curve) with time is called an alternating (AC) source (Fig. 19-5). In this case,

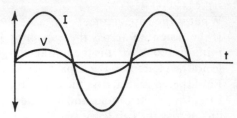

Alternating current and voltage in a purely resistive circuit

Fig. 19-5

Ohm's law takes a slightly different form. Thus 60 cycle, AC has a frequency of 60 Hz (Hertz), or 60 cps and a radian frequency ω of $2\pi 60 = 120\,\pi$ rad/sec. An AC voltage source is,

$$V(t) = V_0 \sin(\omega t + \phi), \quad (19\text{-}4)$$

where $\omega = 2\pi f$ and ϕ is the phase angle.

An inductor in an AC circuit offers an **inductive reactance** X_L to the circuit, where,

$$X_L = 2\pi f L = \omega L, \quad (19\text{-}5)$$

This means that if there are no resistors or capacitors in the circuit, $V = IX_L$. (Compare with $V = IR$.) Furthermore, the current *lags* the voltage (The voltage leads the current.) by 1/4 period; that is, the phase of I is 90° less than the phase of V; thus,

$$I = \frac{V_0}{\omega L} \sin\left(\omega t - \frac{\pi}{2}\right) = I_0 \sin\left(\omega t - \frac{\pi}{2}\right). \quad (19\text{-}6)$$

A capacitor in an AC circuit offers a **capacitive reactance** X_C to the circuit, where,

$$X_C = \frac{1}{\omega C} = \frac{1}{2\pi f C}. \quad (19\text{-}7)$$

The current in a circuit with only a capacitor is $I = V/X_C$. (Compare with $I = V/R$.) In this case, the current *leads* the voltage by 90° and reaches its maximum value $1/4f$ seconds before the voltage. Therefore, V and I are 90° *out of phase*.

(Continued on page 76)

EXPLANATIONS

1. a $\quad C = \dfrac{\epsilon_0 S}{d} = \dfrac{(8.9 \times 10^{-12}\,\text{farad/m})(1\,\text{m}^2)}{10^{-3}\,\text{m}}$

$\quad = 8.9 \times 10^{-9}\,\text{farad} = 8.9 \times 10^{-3}\,\mu\text{f}.$

b $\quad q_0 = CV = (8.9 \times 10^{-9}\,\text{farad})(10^2\,\text{volts})$

$\quad = 8.9 \times 10^{-7}\,\text{coul}.$

c $\quad q(t) = q_0\,\text{Exp}\,(-t/RC):$

$\quad RC = (2 \times 10^3\,\Omega)(8.9 \times 10^{-9}\,\text{farad})$

$\quad = 17.8 \times 10^{-6}\,\text{sec}.$

Since $\quad \dfrac{1}{RC} = \dfrac{1}{17.8 \times 10^{-6}} = 5.6 \times 10^4,$

$\quad q(t) = (8.9 \times 10^{-7})\;\text{Exp}\,(-5.6 \times 10^4 t)\,\text{coul}.$

d $\quad V(t) = \dfrac{q}{C} = \dfrac{(8.9 \times 10^{-7})(\text{Exp}\,(-5.6 \times 10^4 t)\,\text{coul}}{8.9 \times 10^{-9}\,\text{farads}}$

$\quad = 100\,\text{Exp}\,(-5.6 \times 10^4 t)\,\text{volts}.$

e $\quad E = \dfrac{V}{d} = \dfrac{10^2\,\text{Exp}\,(-5.6 \times 10^4 t)\,\text{volts}}{10^{-3}\,\text{m}}$

$\quad = 10^5\,\text{Exp}\,(-5.6 \times 10^4 t)\,\text{volts/m}.$

$\quad \Phi = \mathbf{E} \cdot \mathbf{S} = [10^5\,\text{Exp}\,(-5.6 \times 10^4 t)\,\text{volts/m}]\,(1.0\,\text{m})$

$\quad = 10^5\,\text{Exp}\,(-5.6 \times 10^4 t)\,\text{volts}.$

f. $\quad I_d = \dfrac{\epsilon_0\,d\Phi}{dt}$

$\quad = (8.85 \times 10^{-12}\,\text{farad/m})$

$\quad (-5.6 \times 10^9)\,\text{Exp}\,(-5.6 \times 10^4 t)\,\text{volt m/sec}$

$\quad = (-49.5 \times 10^{-3})[\text{Exp}\,(-5.6 \times 10^4 t)]\,\text{amp}.$

g $\quad \mathbf{B} \cdot \mathbf{l} = B\,2\pi r = \mu_0 I_d$ since $I = 0$ in the capacitor;

hence, $B = \dfrac{\mu_0 I_d}{2\pi r}$

$\quad = \dfrac{\mu_0\,(49.5 \times 10^{-3})\,\text{Exp}\,(-5.6 \times 10^5 t)\,\text{amp}}{(2\pi r)\text{m}}$

$\quad = \dfrac{(9.9 \times 10^{-9})\,\text{Exp}\,(-5.6 \times 10^5 t)}{r}\,\text{weber/m},$

where $\mu_0 = 4\pi \times 10^{-7}\,\text{weber/amp m}.$

but $r = \dfrac{1}{\sqrt{\pi}}$ m at the edge of the capacitor; therefore, $B = (1.75 \times 10^{-9})\,\text{Exp}\,(-5.6 \times 10^5 t)\,\text{weber/m}^2$ at the edge of the capacitor.

h $\quad Z = \sqrt{X_C^2 + R_C^2};$

$\quad X_C = \dfrac{1}{2\pi f C} = \dfrac{1}{2(3.14)(2 \times 10^4\,\text{Hz})(8.85 \times 10^{-9}\,\text{farad})}$

$\quad = 9.0 \times 10^{-2}\,\Omega;$

Answers

$8.9 \times 10^{-3}\,\mu\text{f}$	1 a
$8.9 \times 10^{-7}\,\text{coul}$	b
$(8.9 \times 10^{-7})\,\text{Exp}\,(-5.6 \times 10^4 t)\,\text{coul}$	c
$100\,\text{Exp}\,(-5.6 \times 10^4 t)\,\text{volts}$	d
$10^5\,\text{Exp}\,(-5.6 \times 10^4 t)\,\text{volts}$	e
$(-49.5 \times 10^{-3})\,[\text{Exp}\,(-5.6 \times 10^4 t]\,\text{amp}$	f
$(1.75 \times 10^{-9})\,\text{Exp}\,(-5.6 \times 10^5 t)$	
weber/m^2	g
$2 \times 10^3\,\Omega,\,0.15\,\text{amp, and }29°.$	h
$10^2\,\text{sec}^{-1}$	2 a
in 0.35 sec	b

therefore,

$\quad Z = \sqrt{(81 \times 10^{-4}) + (4 \times 10^6)}\,\Omega$

$\quad = \sqrt{4 \times 10^6}\,\Omega = 2 \times 10^3\,\Omega$ and

$\quad I = \dfrac{V}{Z} = \dfrac{3 \times 10^2\,\text{V}}{2 \times 10^3\,\Omega} = 0.15\,\text{amp}.$

Then

$\quad \tan\phi = \dfrac{X_C}{R} = \dfrac{9 \times 10^{-2}}{2 \times 10^3}\,\Omega = 4.5\,\Omega;$

hence, $\phi \cong 29°.$

2. a $\quad \omega' = \sqrt{\dfrac{1}{(LC)} - \left(\dfrac{R}{2L}\right)^2};$

$\quad L = 5$ henries and

$\quad C = 20\,\mu\text{fd} = 2 \times 10^{-5}$ farad, and $R = 20\;\Omega.$ Thus,

$\quad \omega' = \sqrt{10^4\,\text{sec}^{-2} - 4\,\text{sec}^{-2}}$

$\quad \cong 10^2\,\text{sec}^{-1}$, since the 4 sec^{-2} is negligible.

b The amplitude of I decreases because of the time-dependent factor $\text{Exp}\,(-kt)$; thus $\text{Exp}\,(-kt) = 1/2$ when $kt = 0.694$, therefore $t = 0.694/2 = 0.35$ sec.

resistance. At $t = 0$ the capacitor is charged with q_0 and $I = 0$; the switch is closed and current flows through the circuit. The charge on the capacitor is

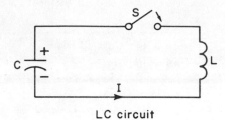

LC circuit

Fig. 19-4

oscillatory and again depends on time,

$$q(t) = q_0 \cos \omega_0 t, \qquad (19\text{-}1a)$$

where $\omega_0 = 1/\sqrt{LC}$, which is the *natural angular frequency*, L is in henries, C is in farads, and ω is in radians/sec (rad/sec). The current $I = dQ/dt$ must oscillate, hence

$$I(t) = -q_0 \omega \sin \omega t = -I_0 \sin \omega t, \quad (19\text{-}2a)$$

where $I_0 = q_0 \omega$ is the maximum value of I. The energy is not dissipated because no resistance is present and the charge decreases *exponentially* with time.

The **RLC circuit** in Fig. 19-2 has an initially charged capacitor. At $t = 0$, the switch S is closed. The current also varies *exponentially* with time; thus,

$$I(t) = I_0 \operatorname{Exp}(-kt) \sin \omega' t, \quad (19\text{-}2b)$$

where $k = R/2L$, and $\omega' = \sqrt{(1/LC) - k^2}$. Eq. 19-2b holds only if R is not too large. Note that $I(t)$ both oscillates and decays with time. This is known as *damped oscillation*. A **driven RLC circuit** is supplied with an alternating source of emf; here, $\epsilon = \epsilon_0 \sin \omega t$. When the emf is applied to the circuit, oscillation will occur as it does in both source-free and driven circuits. After some time has elapsed the current oscillates at the driving frequency ω. The current's amplitude I_0 is,

$$I_0 = \frac{\epsilon_0 \omega / L}{\sqrt{(\omega^2 - \omega_0^2)^2 - k^2}}, \qquad (19\text{-}3)$$

The current will have maximum amplitude when $\omega = \omega_0$, the **resonant frequency.**

A resistor in an AC circuit behaves like a resistor in a DC circuit, thus Ohm's law holds, and V and I are *in phase*. If a circuit contains resistance R plus capacitance C, and/or inductance L, a new quantity called the **impedance** Z is defined, where $I = V/Z$,

$$Z = \sqrt{R^2 + (X_L - X_C)^2} \quad \text{(Fig. 19-6).} \quad (19\text{-}8)$$

The phase angle ϕ between the current and voltage is $\phi = \arctan [(X_L - X_C)/R]$.

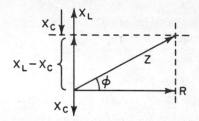

Calculating the impedance using vectors

Fig. 19-6

Ampere's law (Eq. 17-13) can be generalized as follows,

$$Bl = \mu_0 \left(I + \epsilon_0 \frac{\Delta \Phi}{\Delta t} \right). \qquad (19\text{-}9)$$

Therefore, if the electric flux Φ through a surface is changing with time, a magnetic field B is induced in the path l around that surface. The expression $\epsilon_0 \Delta \Phi / \Delta t$ is the **displacement current** I_d. If the electric field E is not changing linearly with time, $I_d = \epsilon_0 d\Phi/dt$. Thus, the modified form of Ampere's law is,

$$\mathbf{B} \cdot \mathbf{l} = \mu_0 (I + I_d), \qquad (19\text{-}9a)$$

or using the calculus,

$$\int \mathbf{B} \cdot d\mathbf{l} = \mu_0 \left(I + \epsilon_0 \frac{d\Phi}{dt} \right) = \mu_0 (I + I_d). \quad (19\text{-}9b)$$

20

LIGHT WAVES: REFLECTION AND REFRACTION

SELF-TEST

1. A man 6 ft tall stands 4 ft from a plane mirror. His eyes are 4 in. below the top of his head. The mirror is just the right length so that he can see his entire height. How long is the mirror?

2. An object is placed 3 cm from a convex lens with a focal length of 6 cm.
   a   Where will the image be formed?
   b   Will the image be real or virtual?
   c   If the object is 1 cm tall, how tall will its image be?

3. A concave mirror is illuminated with light parallel to the axis. Where will the image be formed?

4. A light bulb is placed at the focal point of a convex lens.
   a   Where is the image formed?
   b   If the light bulb is placed at the focal point of a concave lens, where will the image be formed?

5. A ray of light in air strikes a surface of glass at 30° from the normal. Find the angle which the ray makes with the normal inside the glass. (The index of refraction n for the glass is 1.5.)

6. Find the critical angle for a water-air interface. ($n = 1.33$ for water.)

7. A convex lens is made of glass for which $n = 1.40$. The two radii of curvature are 10 cm and 20 cm. What is the focal length of the lens?

8. A candle is placed 10 cm away from the surface of a silvered glass sphere. The radius of the sphere is 20 cm. Draw a ray diagram to locate the image and calculate the position of the image. Is the image real or virtual?

9. A concave mirror has a 20 cm radius of curvature. A light bulb is placed 40 cm to the left of the mirror.
   a   Where will the image be?
   b   How tall will the image be if the diameter of the bulb is 2.0 cm?

1	
2	a
	b
	c
3	
4	a
	b
5	
6	
7	
8	
9	a
	b

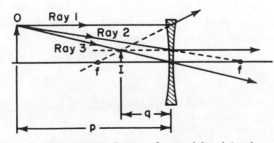

Concave lens; image formed is virtual.

Fig. 20-6

77

BASIC FACTS

A **wavefront** can be thought of as a locus of points having the *same* phase of vibration; for example, the concentric waves caused by dropping a pebble into still water are wavefronts. Thus, all of the points along any wavecrest (maximum) or trough (minimum) have the same phase.

Huygen's principle, which applies to all wave phenomena, states that every point on a three-dimensional wavefront acts as the point source for secondary waves. A point source produces a spherical wave which travels out from the point at velocity c (Fig. 20-1). At a time t seconds

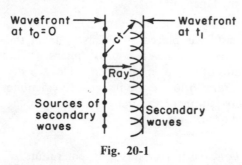

Fig. 20-1

later, the new wavefront will be the surface tangent to the secondary waves. In Fig. 20-1, the wavefront progresses to the right. A wavefront will *refract* (bend) if it passes into a medium for which the velocity of light is slower. Although the path of wavefronts can be deduced from Huygen's principle, it is usually easier to treat only the *ray*, or *normal* to the wavefront. Rays travel in straight lines until they encounter an interface. At that point they are *reflected* and *refracted*. A reflected ray as in Fig. 20-2

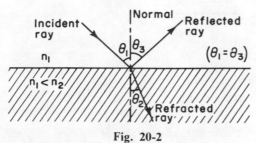

Fig. 20-2

(Continued on page 80)

ADDITIONAL INFORMATION

A **mirror**, whether flat or curved, reflects light rays. When the rays intersect at a point, the point is the location of a **real image**. When the reflected rays do not converge to form an image, but can be extended backwards to meet at a point, that point is the location of a **virtual image**. If a mirror is a section of a sphere with a radius of curvature R, the distance between the mirror and the image is found by drawing the **principal rays**.

The mirror in Fig. 20-3 is called a *concave* mirror. The tip of the arrow at O is the object and the tip of

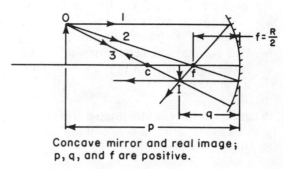

Concave mirror and real image; p, q, and f are positive.

Fig. 20-3

the arrow at I is the image. The three principal rays are as follows:

1. Ray 1, emitted parallel to the axis, is reflected through the **focal point** f on the axis; the distance from f to the mirror is $\frac{1}{2}R$.
2. Ray 2, emitted through f, is reflected parallel to the axis.
3. Ray 3, emitted through point c, the **center of curvature** on the axis, is reflected back onto itself. The distance from point c to the mirror is R, the **radius of curvature**.

The point p is the distance from the mirror to the object, point q is the distance from the mirror to the image, and point c is the distance from the mirror to the center of curvature. If $c = R$, then $f = \frac{1}{2}c$. The formula relating p, q, and f is the **lens equation**,

$$\frac{1}{p} + \frac{1}{q} = \frac{1}{f} = \frac{2}{c} \qquad (20\text{-}2)$$

The following sign conventions are necessary for calculating distances:

1. Points p and q are *positive* if they lie *left* of the mirror, when the mirror is oriented as in Fig. 20-3.

(Continued on page 80)

EXPLANATIONS

1. 2 ft 10 in.

2. a $\dfrac{1}{p} + \dfrac{1}{q} = \dfrac{1}{f}$; $f = 6$ cm, and $p = 3$ cm, thus $q = 6$ cm.

   b Virtual, as it is formed by extended rays.

   c $h_i = h_0 \left(\dfrac{q}{p}\right) = 1\text{ cm} \left(\dfrac{6}{3}\right) = 2$ cm.

3. The image is formed at f since $p = \infty$ (at infinity), thus $\dfrac{1}{p} = 0$; $0 + \dfrac{1}{q} = \dfrac{1}{f}$; therefore $q = f$.

4. a Since $p = f$, $\dfrac{1}{q} + \dfrac{1}{f} = \dfrac{1}{f}$, so that $\dfrac{1}{q} = 0$ and $q = \infty$.

   b $\dfrac{1}{f} + \dfrac{1}{q} = -\dfrac{1}{f}$, so $\dfrac{1}{q} = -\dfrac{2}{f}$, $q = -\frac{1}{2}f$.

5. $\sin \theta_2 = (\sin 30°)\dfrac{n_1}{n_2} = \frac{1}{2}\left(\dfrac{1}{1.5}\right) = 0.33$. $\theta = 18.2°$.

6. $\sin \theta_c = (\sin 90°)\,\dfrac{n_1}{n_2} = \dfrac{1}{1.33} = 0.75$. $\theta_c = 48.5°$.

7. $\dfrac{1}{f} = (n - 1)\left(\dfrac{1}{R_1} + \dfrac{1}{R_2}\right) = (0.40)\left(\dfrac{1}{10} + \dfrac{1}{20}\right)$
   $= 0.060$. $f = 16.7$ cm.

8. $f = -10$ cm, $p = 10$ cm, so $\dfrac{1}{10} + \dfrac{1}{q} = -\dfrac{1}{10}$, $q = -5$ cm
   and the image is virtual.

9. a $f = 10$ cm, $p = 40$ cm, $\dfrac{1}{40} + \dfrac{1}{q} = \dfrac{1}{10}$. $q = \dfrac{40}{3}$
   $= 13.3$ cm.

   b $h_1 = h_0\left(\dfrac{q}{p}\right) = (2\text{ cm})\left(\dfrac{13.3}{40}\text{ cm}\right) = \dfrac{2}{3}$ cm.

Answers	
2 ft 10 in.	1
6 cm	2 a
Virtual	b
2 cm	c
At the focal point	3
At infinity	4 a
At $-\frac{1}{2}f$	b
18.2°	5
48.5°	6
16.7 cm	7
−5 cm, virtual	8
13.3 cm	9 a
$\frac{2}{3}$ cm	b

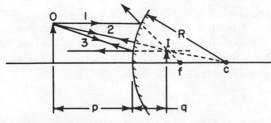

Convex mirror and virtual image;
p is positive, and f and q are negative.

Fig. 20-4

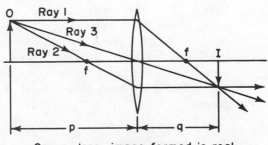

Convex lens; image formed is real.

Fig. 20-5

has two properties: 1. The incident ray, reflected ray, and normal lie in the same plane; 2. the incident angle θ_1 equals the reflected angle θ_3. Notice that θ_1 and θ_3 are both measured from the normal.

Refraction is the bending of a wavefront (and, therefore, of a ray) at the interface between two mediums. The velocity of light is different in the two mediums, and the **index of refraction** n (relative to a vacuum) is therefore also different; n is defined as $n = c/v$, where $c = 3 \times 10^8$ m/sec is the velocity of light in a vacuum and v is the velocity of light in the medium. In Fig. 20-2, $n_1 < n_2$ and the ray is bent *toward* the normal. (Medium 2 is said to be *optically denser* than medium 1.) If the ray passes to an optically less dense medium, the ray is bent *away* from the normal.

Snell's law.   The two angles in Fig. 20-2 are related to n_1 and n_2 as follows:

$$\frac{\sin \theta_1}{\sin \theta_2} = \frac{n_2}{n_1} \qquad (20\text{-}1)$$

Part of a ray is reflected when it crosses a boundary. If $n_1 > n_2$, and θ_1 is greater than the **critical angle** θ_c, the ray is completely reflected. The critical angle is found by setting θ_2 equal to 90° in Eq. 20-1; thus, $\sin \theta_2 = 1$ and $\sin \theta_c/1 = n_2/n_1$. For air, $n = 1.00$, thus if medium 1 is air, Snell's law becomes $\sin \theta_1 / \sin \theta_2 = n_2$.

Light has a lower velocity in a material medium than in a vacuum. The velocities of different wavelengths of light are different from one another, hence n is a function of λ. Thus a beam of white light refracted at an interface breaks up into beams of different component colors. This phenomenon is called **dispersion**.

2. Points f and c are *positive* if they lie to the *left* of the mirror—f and c are *negative* for a *convex* mirror since they lie to the *right* of the mirror.

The mirror in Fig. 20-4 is *convex*. Point p is *positive*, points q and f are *negative*, and the image is *virtual*. The height h_i of the image is related to the height h_0 of the object by,

$$\frac{h_i}{h_0} = \frac{q}{p}. \qquad (20\text{-}3)$$

A thin lens is a piece of transparent material with two radii of curvature, as shown in Fig. 20-5 and Fig. 20-6. The **focal length** is given by the formula called the **lensmaker's equation**,

$$\frac{1}{f} = (n-1)\left(\frac{1}{R_1} + \frac{1}{R_2}\right), \qquad (20\text{-}4)$$

where n is the index of refraction of the material; f is *positive* for the convex (converging) lens shown in Fig. 20-5 and *negative* for the concave (diverging) lens shown in Fig. 20-6. The principal rays for locating the image formed by the refracted rays of a lens are similar to those of a mirror.

In Fig. 20-5, the principal rays are 1, 2, and 3. Ray 1 is parallel to the axis and is refracted through the focal point f on the opposite side of the lens, ray 2 goes through f on the same side as O and is refracted parallel to the axis, and ray 3 passes through the center of the lens and is undeflected. In Fig. 20-6, ray 1 is parallel to the axis and is refracted so that it appears to have come from the focal point on the left. Ray 2 aims at the focal point on the right of the lens and is refracted parallel to the axis. Ray 3, through the center, is undeflected. When rays 1, 2, and 3 are extended backwards, they meet at a point to form a virtual image. As with mirrors, a real image is formed by the intersection of rays and a virtual image is formed by the intersection of extended rays. Eq. 20-2 also applies to lenses, but the sign conventions for p, q, and f are slightly different: For a *convex* lens, f is *positive*. For a *concave* lens, f is *negative*, q is *positive* if I is on the opposite side of the lens from O, I is *real* if q is positive, q is negative if I is on the same side of the lens as O, and I is virtual if q is negative.

21

DIFFRACTION AND INTERFERENCE OF LIGHT

SELF-TEST

1. A light wave of wavelength $\lambda = 5000$ Å shines on a slit of width $4.0 \times 10^{-\cdot}$ cm. The interference pattern is displayed on a screen 50 cm behind the slit (Fig. 21-1).

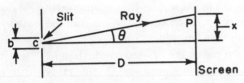

b=length of slit, c is the center of slit, P is the point where the ray hits the screen, and D and x are distances.

Fig. 21-1

a How far is the first minimum from the central maximum (maximum for which $N = 0$, $x = 0$)?

b Approximately how far is the third maximum from the central maximum?

c Find the ratio of the intensity of the third maximum to the intensity of the central maximum.

2. Two slits, each 5.0×10^{-2} cm in width, are separated by 0.25 cm. Light with a wavelength of 6000 Å shines on the slits; a screen 1.00 m behind the slits displays the diffraction pattern.

a How far from the central maximum are the first and second maxima?

b Are there any missing maxima? If so, list a few.

3. A certain diffraction grating consists of 25 slits, 3×10^{-4} cm wide, separated by 6×10^{-3} cm; $\lambda = 3000$ Å. The pattern shines on a screen 2 m away.

a Find the distance between the central maximum and the first principal maximum.

b Which maxima are missing?

4. The radius of the tenth ring in a set of Newton's rings is 0.20 cm. The radius of curvature of the lens is 30 cm. Calculate the wavelength of the light used.

5. A glass lens with an index of refraction $n = 1.50$ is coated with a thin layer of a transparent substance. The purpose of the coating is to eliminate the glare caused by reflected light. Use $\lambda = 5500$ Å, since this is the center of the visible spectrum.

a If $n = 1.40$ for the coating substance, calculate the smallest optimum thickness for this layer.

b If $n = 1.60$ for the coating substance, calculate the smallest optimum thickness for the layer.

1 a
 b
 c

2 a
 b

3 a
 b

4

5 a
 b

BASIC FACTS

Any two light waves add by *superposition*, but in order to produce **interference maxima** and **minima** which last long enough to be observed, both waves must have the same frequency and a constant phase relationship. These requirements are the conditions for *coherence*. The amplitude of the resulting wave is just the sum of the amplitudes of the individual waves, according to the principle of superposition (chapter 13); the intensity or power per unit area of the wave is proportional to the square of the amplitude of the wave. When the phases $2\pi \left(ft + \dfrac{x}{\lambda} \right)$ of two such waves at a point are equal or differ by 2π, 4π, 6π, ... the result is a bright spot or *maximum*. When the phases differ by π, 3π, 5π, ..., the result is a dark spot or *minimum*.

Light shining on a **single slit** in an opaque wall produces a diffraction pattern on a screen placed beyond the wall, because the secondary waves from different portions of the slit travel different distances to reach a point on the screen. Fig. 21-1 gives a side view of a slit of length b, P is a point on the screen, and the intensity of light at P is expressed as $I(\theta)$; thus,

$$I(\theta) = I_o \left(\frac{\sin y}{y} \right)^2, \qquad (21\text{-}1)$$

where, $y = (\pi b \sin \theta)/\lambda$, I_o is the intensity at $\theta = 0$, and λ is the wavelength of the light wave. **Minima** occur when $\sin y = 0$; that is, when $y = N\pi$ radians ($N = 1,2,\ldots$). Solving for $\sin \theta$ and letting $y = N\pi$,

$$\sin \theta = \frac{N\lambda}{b} \quad (\text{min}). \qquad (21\text{-}2)$$

Maxima occur when $\sin y = 1$ (approx); that is, when $y = (N + \frac{1}{2})\pi$ radians ($N = 0,1,2,\ldots$); thus,

$$\sin \theta = \frac{(N + \frac{1}{2})\lambda}{b} \quad (\text{max}). \qquad (21\text{-}2a)$$

(Continued on page 84)

ADDITIONAL INFORMATION

A plane light wave shines on the left side of the barrier in Fig. 21-2. The barrier stops all the light except at the slits. According to Huygen's principle, the light at the slits is the source of secondary waves which radiate from the slits. These waves interfere (add by superposition) at every point P on the screen placed a distance D to the right of the barrier. The distance between two successive *crests* (maximum amplitude) of either wave is one wavelength λ and the distance between a crest and a *trough* (minimum amplitude) of either wave is $\lambda/2$. If the crests or troughs of the waves from the two slits coincide at P, the waves are *in phase* and there is an intensity maximum at the point. If however, the crests of the waves from one slit coincide with the troughs from the other slit, then the waves are 180° out of phase and P will be a point of minimum intensity. The distances from P on the screen to slits s_1 and s_2 are D_1 and D_2, respectively; P is point of maximum intensity if the waves from s_1 and s_2 are in phase at P. Then,

$$D_1 - D_2 = \pm N\lambda \ (\text{max}), N = 0,1,2,\ldots. \qquad (21\text{-}7)$$

P is a point of minimum intensity if,

$$D_1 - D_2 = \pm (N + \tfrac{1}{2})\lambda \ (\text{min}), N = 0,1,2,\ldots. \qquad (21\text{-}7a)$$

If $d \ll D$, then $\theta_1 = \theta_2 = \theta$, and $x/D = \tan \theta$ and $N\lambda/d = \sin \theta$ for maxima. For small angles ($\theta \ll 10°$), $\tan \theta$ is very nearly equal to $\sin \theta$; thus $x/D = N\lambda/d$, so that $N\lambda = d \sin \theta$ and $x = D \sin \theta$; combining these equations,

$$\sin \theta = \frac{N\lambda}{d} \quad (\text{max}), N = 0,1,2,\ldots, \qquad (21\text{-}8)$$

and, $$x = \frac{DN\lambda}{d}. \qquad (21\text{-}8a)$$

A light wave shining on an interface (surface between two media) is divided into two waves (chapter 20): a *reflected* wave and a *refracted* wave. A **thin film** has two closely spaced *interfaces* (Fig. 21-3). Ray 1 splits into rays 2 and 3, and ray 3 splits into rays 4 and 6. Parallel rays 2 and 5 are made to interfere on the screen by the lens placed one focal length f from the screen. The phases of rays 2 and 5 are different for two reasons: the path of rays 3, 4, and 5 is longer than that of ray 2, and the phase of ray 2 changes by

(Continued on page 84)

EXPLANATIONS

1. a $N = 1$ and $\sin \theta = \dfrac{N\lambda}{b} = \dfrac{x}{D}$, where $N = 1$; therefore,

$$x = \frac{(0.5 \text{ m})(1)(5 \times 10^{-7} \text{ m})}{4 \times 10^{-4} \text{ m}} = 6.25 \times 10^{-4} \text{ m}.$$

   b $N = 3$ and $x = \dfrac{(0.5 \text{ m})(7/2)(5 \times 10^{-7} \text{ m})}{4 \times 10^{-4} \text{ m}} = 2.2 \times$

10^{-3} m.

   c $\dfrac{I_3}{I_0} = \left(\dfrac{\sin y}{y}\right)^2$ (Eq. 21-1) and $y = \dfrac{\pi a \sin \theta}{\lambda} = \pi(N + \tfrac{1}{2})$

for a maximum by Eq. 21-2a; thus, $\dfrac{I_3}{I_0} = \left(\dfrac{\sin \frac{7}{2}\pi}{\frac{7}{2}\pi}\right)^2 = $

$-\left(\dfrac{1}{\frac{7}{2}\pi}\right)^2 = \dfrac{4}{49\pi^2} = 8.31 \times 10^{-3}$.

2. a $\sin \theta = \dfrac{x}{D} = \dfrac{N\lambda}{d}$, therefore $x_1 = \dfrac{DN\lambda}{d}$

$$= \frac{(1 \text{ m})(1)(6 \times 10^{-7} \text{ m})}{2.5 \times 10^{-3} \text{ m}} = 2.4 \times 10^{-4} \text{ m}.$$

   b Missing maxima for $\dfrac{b}{d} = \dfrac{N}{M}$, where $\dfrac{b}{d} = \dfrac{1}{5}$; thus, $M = 5N$ is the condition. The 5th, 10th, 15th, ... maxima are missing.

3. a $\sin \theta = \dfrac{M\lambda}{d} = \dfrac{x}{D}$; $x = \dfrac{(2 \text{ m})(1)(3 \times 10^{-7} \text{ m})}{6 \times 10^{-5} \text{ m}} = $

10^{-4} m.

   b $\dfrac{b}{d} = \dfrac{3 \times 10^{-4}}{6 \times 10^{-3}} = \dfrac{1 \times 10^{-1}}{2} = \dfrac{1}{20} = \dfrac{N}{M}$; thus, $M = 20N$
and the missing maxima are the 20th, 40th, 60th, ...

4. $d = \dfrac{r^2}{2R} = \dfrac{(2 \times 10^{-3} \text{ m})^2}{2 \times 0.3 \text{ m}} = 0.67 \times 10^{-5}$ m and $2d = $

$(N + \tfrac{1}{2})\lambda$; thus, $\lambda = \dfrac{2d}{N + \frac{1}{2}} = \dfrac{2(0.67 \times 10^{-5} \text{ m})}{10 + \frac{1}{2}} = 1.27 \times$

10^{-6} m.

5. a $N\lambda = 2dn_2$ and $N = 1$ for the smallest optimum thickness. $d = \dfrac{\lambda N}{2n_2} = \dfrac{(5500 \text{ Å})(1)}{2(1.40)} = 1970 \text{ Å}.$

   b The refractive index for the coating substance is now greater than that for glass and $N = 0$ for the smallest optimum thickness, thus $(N + \tfrac{1}{2})\lambda = 2dn_2$ and $d = \dfrac{(N + \frac{1}{2})\lambda}{2n_2} = \dfrac{(0 + \frac{1}{2})(5500 \text{ Å})}{2(1.60)} = \dfrac{5500 \text{ Å}}{6.40} = 860 \text{ Å}.$

Answers

6.25×10^{-4} m	1 a
2.2×10^{-3} m	b
8.31×10^{-3}	c
2.4×10^{-4} m	2 a
5th, 10th, 15th, ...	b
10^{-4} m	3 a
20th, 40th, 60th, ...	b
1.27×10^{-6} m	4
1970 Å	5 a
860 Å	b

For a **double slit**, the waves from each slit are *coherent* and the rays from each slit travel to point P on the screen (Fig. 21-2). In this case,

$$I(\theta) = I_0 \left(\frac{\sin y}{y}\right)^2 \cos^2 z, \qquad (21\text{-}3)$$

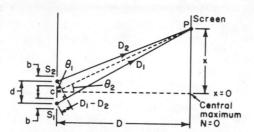

S_2 = slit 2 and S_1 = slit l, b is the length of the slits, d is the distance between the centers of the slits, and c is the center between the slits.
$\lambda/d = \sin\theta$, and $x \cong D\sin\theta$.

Fig. 21-2

where $z = (\pi d \sin\theta)/\lambda$; d is the distance between the slits. The intensity is zero whenever Eq. 21-2 is satisfied; it is also zero whenever $z = (M + \frac{1}{2})\pi$ ($M = 0,1,2,\ldots$), or when,

$$\sin\theta = \frac{(M + \frac{1}{2})\lambda}{d} \quad (\text{min}) \qquad (21\text{-}4)$$

Intensity maxima occur whenever $z = M$ ($M = 0,1,2,\ldots$); thus,

$$\sin\theta = \frac{M\lambda}{d} \quad (\text{max}) \qquad (21\text{-}4a)$$

If Eq. 21-2 and 21-4b both hold, P will be a minimum. This is the condition for a missing maximum; equating the right-hand sides of Eq. 21-2 and 21-4b,

$$\frac{d}{b} = \frac{M}{N} \qquad (21\text{-}5)$$

A number of parallel slits is called a **diffraction grating**. Each slit has a width b and is separated by a distance d; then,

$$I(\theta) = I_0 \left(\frac{\sin y}{y}\right)^2 \left(\frac{\sin nz}{\sin z}\right)^2, \qquad (21\text{-}6)$$

where n is the number of slits. The conditions for minima, maxima, and missing maxima are the same as those in Eq. 21-4, 21-4b, and 21-5, respectively.

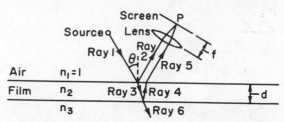

n_1, n_2, and n_3 are the indices of refraction, l, 2, 3, 4, and 5 are the rays, f is the focal length of the lens, and d is the thickness of the film.

Fig. 21-3

π radians (180°) from the phase of ray 1. In fact, whenever a wave in a medium of index n_1 reflects from a medium of index $n_2 > n_1$, the wave undergoes a phase change of 180°. Ray 3 does not undergo a phase change whenever $n_3 < n_2$.

The conditions for interference (when $\theta = 0$) are the following:

1. If $n_2 > n_3$, $N\lambda = 2dn_2$ (min), and $(N + \frac{1}{2})\lambda = 2dn_2$ (max), $N = 0,1,2,\ldots$.
2. If $n_2 < n_3$, the conditions are reversed: $(N + \frac{1}{2})\lambda = 2dn_2$ (min), and $N\lambda = 2dn_2$ (max).

Newton's rings, or circular interference fringes (concentric, alternating bright and dark areas) occur when a *convex* glass lens placed on a flat glass plate (Fig. 21-4) is illuminated by light; R is the *radius of curva-*

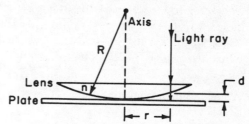

Light rays incident on the lens are parallel to the axis; $d = \frac{r^2}{2R}$.

Fig. 21-4

ture of one face of the lens; the other face is flat. Interference occurs because light incident on the lens at different values of r travels different distances d through air; d for the lens is,

$$d = \frac{r^2}{2R}. \qquad (21\text{-}9)$$

If light shines down parallel to the axis of the lens, the conditions for maximum and minimum interference are,

$$2d = \frac{r^2}{R} = (N + \frac{1}{2})\lambda \quad (\text{max}), \qquad (21\text{-}10)$$

and, $\quad 2d = \frac{r^2}{R} = N\lambda \quad (\text{min}), \qquad (21\text{-}10a)$

where the index of refraction of air is unity and $N = 0,1,2,\ldots$

22

RELATIVITY

1. If each day the earth receives about 1.5×10^{22} joule of energy from the sun in the form of electromagnetic radiation and if the earth does not reradiate most of this energy (which it does), by what amount would the mass of the earth increase each day? (The speed of light $c = 3 \times 10^8$ m/sec.)

2. An electron has a rest mass of 9.1×10^{-31} kg. In a certain experiment, the electron's kinetic energy is 25.6×10^{-7} joule.
   a What is the rest energy of the electron?
   b What is the total energy of the electron?
   c What is the momentum of the electron?

3. A man on a spaceship flying past the earth at a velocity of $0.99c$ (c = the speed of light) relative to the earth finds that he can read 500 words per minute. How many words per minute can the man read when he is timed by clocks on earth?

4. A neutron emitted from a star has a kinetic energy of 1.5×10^{-7} joule. It decays in 12 minutes (measured in its own frame) into an electron and a proton. The rest mass of a neutron is 1.67×10^{-27} kg.
   a What is the rest energy of the neutron and how does it compare with the kinetic energy?
   b Find the neutron's total energy.
   c Find the neutron's momentum.
   d How long does the neutron travel in the star's frame before it decays?
   e How far does the neutron go in its own frame?
   f If an observer in the star's frame measures the equatorial diameter of the earth and finds it to be about 8000 miles, what would an observer in the neutron's frame see?

5. Light emitted by a source 200 miles away from an observer has a frequency of 3×10^{15} Hz (cycle/sec). An observer measures the light frequency to be 9×10^{15} Hz. Explain.

1	
2	a
	b
	c
3	
4	a
	b
	c
	d
	e
	f
5	

BASIC FACTS

All physical quantities are measured relative to a *reference frame*. In chapter 8, it was shown that if the frame is not inertial, ficticious forces arise; however, in inertial frames no such ficticious forces appear. The following **postulates of special relativity** apply to inertial frames.

1. All the laws of physics retain the same form in all inertial frames.
2. The speed of light c ($c = 3 \times 10^8$ m/sec) is constant in all inertial frames.

Postulate 1 implies that length and time measurements in one frame can be transformed linearly into length and time measurements in another frame. The **Lorentz transformations** perform this function. Letting $\gamma = 1/\sqrt{1 - v^2/c^2}$, these transformations are,

$$x' = (x - vt)\gamma, \qquad (22\text{-}1)$$

and,

$$t' = (x - vx/c^2)\gamma, \qquad (22\text{-}2)$$

where the prime (') denotes measurement in the moving frame, x' = distance, and t' = time. The primed frame will henceforth be designated by O' and the unprimed frame by O. In Fig. 22-1, the two

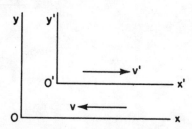

v' is the velocity of O' relative to O;
v is the velocity of O relative to O'.

Fig. 22-1

reference frames are represented by coordinate axes xy and $x'y'$; here, O and O' are the origins of the stationary and moving frames, respectively. The x' and t' on the O' frame can be transformed into x and t in O frame, since as far as

(Continued on page 88)

ADDITIONAL INFORMATION

Any energy W, be it chemical, thermal, potential, or nuclear has a mass associated with it given by,

$$m = \frac{W}{c^2}. \qquad \text{(See Eq. 22-7a.)}$$

Example 1. A 1 kg block of ice at 0° C is melted to form water. What will the increase of the mass of the water be? In order to melt the ice, 80 kcal of heat must be supplied. Since 1 kcal = 4190 joule, 3.35×10^5 joule must be added to the ice. The increase in mass m is therefore,

$$m = \frac{W}{c^2} = \frac{3.35 \times 10^5 \, \text{joule}}{(3 \times 10^8 \, \text{m/sec})^2}$$

$$= 3.72 \times 10^{-12} \text{ kg, which is not very much.}$$

The positions and times of the two reference frames O and O' in Fig. 22-1 are transformed whenever $v \ll c$ by the **nonrelativistic**, or **Galilean transformation**,

$$x' = x - vt, \qquad (22\text{-}1b)$$

$$x = x' + vt', \qquad (22\text{-}1c)$$

and,

$$t = t'. \qquad (22\text{-}2b)$$

If $v \ll c$, then the above equations are the same as Eq. 22-1, Eq. 22-1a, Eq. 22-2 and Eq. 22-2a, respectively. Eq. 22-2b holds because the same time scale applies to all reference systems.

Lengths, time intervals, masses, momenta, and energies behave for low speeds as described in previous chapters.

Example 2. The mass of an object traveling at a high velocity is twice that of the stationary object.

a. If the object's length is L_0 when it is stationary, what will be its length when it is traveling at velocity v?
b. Find v.

a. Solving Eq. 22-5 for γ, $m = \frac{1}{2}m\gamma$, thus $\gamma = 2$; from Eq. 22-3, $L = L_0/2$. Note that m_0 and L_0 are substituted for m' and L', respectively.
b. To find v, use $\gamma = 2 = 1/\sqrt{1 - v^2/c^2}$; therefore, $v = \sqrt{(3/4)c^2}$. Noting that the constant for the speed of light $c = 3 \times 10^8$ m/sec, then $v = \sqrt{(3/4)(3 \times 10^8 \, \text{m/sec})^2} = 2.6 \times 10^8$ m/sec.

(Continued on page 88)

EXPLANATIONS

1. $m = \dfrac{W}{c^2} = \dfrac{(1.5 \times 10^{22}\text{ joule})}{(3 \times 10^8\text{ m/sec})^2} = 1.67 \times 10^5$ kg.

2. a $W_0 = m_0 c^2 = (9.1 \times 10^{-31}$ kg$)(3 \times 10^8$ m/sec$)^2 = 8.2 \times 10^{-15}$ joule.

   b $W = K + W_0 = (25.6 + 8.2) \times 10^{-15}$ joule $= 33.8 \times 10^{-15}$ joule.

   c $p^2 c^2 = W^2 - W_0^2 = (33.8 \times 10^{-15})^2 - (8.2 \times 10^{-15}$ joule$)^2 = 1075 \times 10^{-30}$ joule;

   thus, $p = \dfrac{\sqrt{1075} \times 10^{-15}\text{ joule}}{3 \times 10^8\text{ m/sec}}$

   $= 1.02 \times 10^{-22}$ kg m/sec.

3. $t' = 1$ min on spaceship and $t = t'\gamma$ on earth; but $v = 0.99c$, hence $t = (1\text{ min})\gamma = \dfrac{1\text{ min}}{\sqrt{1 - 0.98}} = 7.07$ min. Therefore, the man can read $(500\text{ words})/(7.07\text{ min}) = 70.7$ words/min.

4. a $W_0 = m_0 c^2 = (1.67 \times 10^{-27}$ kg$)(3 \times 10^8$ m/sec$)^2 = 1.5 \times 10^{-10}$ joule, which is small compared with the kinetic energy.

   b $W = K + W_0 \cong K = 1.5 \times 10^{-7}$ joule, since $K \gg W_0$.

   c Since $p^2 c^2 \cong K^2$, $p \cong \dfrac{K}{c} = \dfrac{1.5 \times 10^{-7}\text{ joule}}{3 \times 10^8\text{ m/sec}} = 5 \times 10^{-16}$ kg m/sec.

   d $t = t'\gamma = (12\text{ min})\gamma$, since $t' = 12$ min in the electron's frame. Now $\gamma = \dfrac{W}{W_0} = \dfrac{1.5 \times 10^{-7}\text{ joule}}{1.5 \times 10^{-10}\text{ joule}} = 10^3$, thus $t = 12 \times 10^3$ min.

   e $v \cong c$, so that the distance $x = ct = (3 \times 10^8$ m/sec$)(7.2 \times 10^5$ sec$) = 21.6 \times 10^{13}$ m in the star's frame. This distance is contracted from the point of view of the neutron; thus,

   $$x' = \frac{x}{\gamma} = \frac{21.6 \times 10^{13}\text{ m}}{10^3} = 2.16 \times 10^{11}\text{ m}.$$

   f The 8000 mile diameter is contracted in the direction of the motion of the neutron. From Eq. 22-3, letting $L' = D'$, then $D' = \dfrac{D}{\gamma} = \dfrac{8000\text{ mi}}{10^3} = 8$ mi. The earth flattens out to a disk 8000 mi in diameter and 8 mi thick.

5. From Eq. 22-11,

   $$\frac{v}{c} = \frac{f_s^2 - f_0^2}{f_s^2 + f_0^2} = \frac{(9 - 81) \times 10^{30}\text{ Hz}}{(9 + 81) \times 10^{30}\text{ Hz}} = -0.8,$$

   where the velocity is toward the observer.

an observer on O' is concerned, O is moving past him with velocity $-v$ (in the opposite direction). This transformation is given by,

$$x = (x' + vt')\gamma, \qquad (22\text{-}1a)$$

and, $\qquad t = (t' + vx/c^2)\gamma. \qquad (22\text{-}2a)$

The length of an object in O' is L', when measured by someone in O'. When measured by someone in O, it is L, where,

$$L = L'/\gamma. \qquad (22\text{-}3)$$

Notice that the length of a moving object contracts.

A time interval measured in O' is t'; the same time interval measured in O is t, where,

$$t = t'\gamma. \qquad (22\text{-}4)$$

Notice that the time interval in a moving frame dilates (gets bigger).

The mass of an object moving with O' is m'; the same mass measured by an observer moving with O is m; thus,

$$m = m'\gamma, \qquad (22\text{-}5)$$

where m' is usually denoted by m_0 and is referred to as the **rest mass**.

The momentum p in relativistic mechanics is given by,

$$p = mv = m_0\gamma v. \qquad (22\text{-}6)$$

A particle has an energy even when it has no potential or kinetic energy. This energy is called the **rest energy** W_0, where,

$$W_0 = m_0 c^2. \qquad (22\text{-}7)$$

The total energy, kinetic energy plus rest energy, of a particle is,

$$W = mc^2 = m_0\gamma c^2. \qquad (22\text{-}7a)$$

The kinetic energy K is the difference between Eq. 22-7 and Eq. 22-7a; thus,

$$K = W - W_0 = (m - m_0)c^2. \qquad (22\text{-}8)$$

Momentum p in classical mechanics is related to the kinetic energy by $K = p^2/2m$; but, in relativistic mechanics,

$$p^2 c^2 = W^2 - W_0^2 \qquad (22\text{-}9)$$
$$= (m^2 - m_0^2)c^4.$$

Because the *Doppler shift* (chapter 13) for sound waves depends on the velocity of the observer and the source relative to the medium (air), it can be determined whether the observer or the source is moving relative to the air. However, the Doppler shift for a light beam depends only on the relative velocities of the observer and the source, but not on the medium through which it passes. The frequency f_0 of light detected by the observer is related to the frequency f_s of the source by the following equation,

$$f_0 = f_s \sqrt{\frac{1 - v/c}{1 + v/c}}. \qquad (22\text{-}10)$$

A very good approximation of Eq. 22-11 is given by,

$$f_0 = f_s \frac{c - v}{c}, \qquad (22\text{-}10a)$$

when $v \ll c$, since, in this case, $(1 + v/c)^{-1}$ can be set equal to $1 - v/c$.

The velocity of an object measured in O and O' is u and u', respectively. The relativistic formulas relating u and u' are,

$$u = \frac{u' + v}{1 + u'v/c^2}, \qquad (22\text{-}11)$$

and, $\qquad u' = \dfrac{u - v}{1 + uv/c^2}, \qquad (22\text{-}11a)$

if u is parallel to v.

Final Examination
Dictionary-Index

FINAL EXAMINATION

Questions

DIRECTIONS: Write your answers on the numbered lines in the margins. To check your answers, turn to page 92. Study the explanations for any questions you have missed.

Answers
1a
b
c
d

2a
b
c
d

3

4

5a
b

6a
b

7a

b

8

9a
b
c

10a

b

1. A ball is thrown at an angle of 60° from the horizontal; its initial velocity is 100 ft/sec.
   a How high does the ball rise?
   b How long does it take the ball to reach this height?
   c How long is the ball in the air, if it falls to the same height from which it is thrown?
   d How far does the ball travel horizontally before hitting the ground?

2. A man standing at the edge of a cliff shoots a gun pointed at an angle of 30° from the horizontal. The initial velocity of the bullet is 1000 ft/sec and the height of the cliff is 500 ft.
   a Find the maximum height the bullet rises above the cliff.
   b Find the bullet's time of flight t_f.
   c Find the x-distance traveled by the bullet.
   d Find the bullet's final velocity.

3. A 1500 kg truck traveling north at 20 m/sec collides with a 2000 kg truck traveling east at 20 m/sec; the collision is inelastic. What is the velocity of the combined wreck just after the collision?

4. A uniform, 4 kg, metal sheet 5 × 10 cm is free to rotate about an axis in the plane of the sheet, through the center of mass, and parallel to the short edge. What is the moment of inertia of the sheet?

5. A 160 lb man stands on a scale on a cart which rolls frictionlessly down an incline. During the descent, the man reads 120 lb on the scale.
   a What is the vertical component of the acceleration of the cart?
   b What is the angle θ of the incline?

6. A certain planet has a mass of $(1/6.67) \times 10^{21}$ kg and a radius of 2×10^4 m.
   a Calculate the acceleration of gravity g at the surface of the planet.
   b Calculate the potential energy of a 2 kg mass at the surface of the planet relative to zero potential energy at infinity.

7. A cubic block, 1 m on each edge, is placed in a liquid having a density of 2.04×10^3 kg/m^3. The cube floats with 0.8 m of the vertical edge submerged.
   a What is the density of the block?
   b If the mass is pushed down a distance x below the equilibrium point and released, what is its acceleration?

8. How much longitudinal pressure must be applied to fit a steel rod 2 cm × 2 cm × 3 m into a container 2 cm × 2 cm × 2.94 m? Young's modulus for steel is 2.0×10^{12} dy/cm^2.

9. 60 gm of O_2 gas is heated at a constant atmospheric pressure (10 nt/m^2) from 27.0° C to 127.0° C. $C_v = 5.03$ cal/mole° K.
   a What is the change in internal energy of the gas?
   b How much heat was added to the gas?
   c How much work was done by the gas?

10. A 2 m long rod with a 4 cm^2 cross section is well insulated to prevent heat losses. One end is welded to a steam bath at 100°C and the other end is attached to a water jacket. Circulating water enters the jacket at 19°C and leaves at 21°C when the flow rate is 20 gm/sec.
    a How much heat is transferred down the rod each second?
    b What is the thermal conductivity of the rod?

11. A Carnot engine takes in 2×10^4 joule of heat during the isothermal expansion at $600°$K. If the engine is to perform 10^4 joule of work, what is the upper limit of the lower temperature?

12. A 10 m, 250 gm wire is stretched between two walls. The fundamental frequency of the wire is found to be 350 Hz.
    a   What is the tension in the wire?
    b   If a cold spell causes the tension in the wire to double, what will the fundamental frequency be?

13. A parallel-plate capacitor is connected to a 600 volt battery. The force on a 10^{-4} coul charge placed between the plates is measured to be 3 nt. What is the electric field between the plates?

14. If 10 m is the distance between the wing tips of an airplane flying horizontally at 100 m/sec in a region for which the vertical component of the earth's magnetic field is 2 gauss, what is the potential difference between the wing tips?

15. Calculate the inductance L of a long solenoid that has a constant magnetic field $B = \mu_0 nI$ (Eq. 17-10b), where $n = N/l$ is the number of turns per unit length.

16. Find the critical angle for a water-carbon bisulfide interface. ($n = 1.33$ for water and $n = 1.63$ for carbon bisulfide.)

17. A soap film of $n = 1.5$ on a wire loop is illuminated by white light. The thickness of the film is 3000 Å. What is the color of the light that will be reflected from the film?

18. A convex lens flat on one side is placed curved face down on a reflecting surface. Dark circles are observed when the lens is illuminated from above by light of wavelength 6000 Å. The radius of the 100th dark circle is measured to be 1.2 cm. What is the radius of curvature of the lens?

11

12a
b

13

14

15

16

17

18

Answers

117.2 ft	**1a**
2.71 sec	**b**
5.42 sec	**c**
271 ft	**d**
3906 ft	**2a**
32.2 sec	**b**
2.79×10^4 ft	**c**
-530 ft/sec	**d**
14.3 m/sec	**3**
3.3×10^{-3} kg m^2	**4**
8 ft/sec^2	**5a**
30°	**b**
25 m/sec^2	**6a**
-10^6 joule	**b**
1.63 kg/m^3	**7a**
$-12.25x$ m/sec^2	**b**
4×10^{10} dy/cm^3	**8**
6.24×10^3 joule	**9a**
942 cal	**b**
1315.5 cal	**c**
40 cal/sec	**10a**
+2.5 cal/cm sec^2 °C	**b**

Explanations

1. a $v_{0y} = v_0 \sin 60° = (100 \text{ ft/sec})(0.866) = 86.6$ ft/sec, $v_{0x} = v_0 \cos 60° = (100 \text{ ft/sec})(0.50) = 50$ ft/sec, and $y = y_{max}$ when $v_y = 0$; thus $v_y^2 = v_{0y}^2 - 2gy_{max}$. Therefore, $y_{max} = \dfrac{v_{0y}^2}{2g} = \dfrac{(86.6 \text{ ft/sec})^2}{(2)(32 \text{ ft/sec}^2)} = 117.2$ ft.

b $v_y = v_{0y} - gt = 0$; thus $t = v_{0y}$ and

$$g = \frac{86.6 \text{ ft/sec}}{32 \text{ ft/sec}^2} = 2.71 \text{ sec}.$$

c Since it takes 2.71 sec to fall back, the total time is $t_{tot} = 2 \times 2.71$ sec $= 5.42$ sec. Also, since $y = 0 = v_{0y}t_{tot} - gt^2$,

$$t = \frac{v_0}{\frac{1}{2}g} = \frac{86.6 \text{ ft/sec}}{\frac{1}{2}(32 \text{ ft/sec}^2)} = 5.42 \text{ sec}.$$

d $x = v_{0x}t_{tot} = (50 \text{ ft/sec})(5.42 \text{ sec}) = 271$ ft.

2. a $v_{0x} = v_0 \cos 30°$
   $= (1000 \text{ ft/sec})(0.866) = 866$ ft/sec;
   $v_{0y} = v_0 \sin 30°$
   $= (1000 \text{ ft/sec})(0.50) = 500$ ft/sec.
   $v_y = 0$ at $(y_{max} = h)$; the maximum height; therefore,
   $v_y^2 = 0 = v_{0y}^2 - 2gh$, so that

$$h = \frac{v_{0y}^2}{2g}$$

$$= \frac{(500 \text{ ft/sec})^2}{2 \times 32 \text{ ft/sec}^2} = 3906 \text{ ft}.$$

b $y_{final} = -500 \text{ ft} = v_{0y}t_f - \frac{1}{2}gt^2 =$

$(500 \text{ ft/sec})t_f - \frac{1}{2}(32 \text{ ft/sec}^2)t^2$, then

$t_f^2 - 31.2 \ t_f - 31.2 = 0$ and $t_f = \dfrac{31.2}{2} \pm$

$\frac{1}{2}\sqrt{(31.2)^2 + 4(31.2)}$; the positive root is

used; thus, $t_f = 32.2$ sec.
   c $x = v_{0x}t = (866 \text{ ft/sec})(32.2 \text{ sec})$
   $= 2.79 \times 10^4$ ft.
   d $v_{fx} = v_{0x} = 866$ ft/sec, and $v_{fy} = v_{0y} - gt = (500 \text{ ft/sec}) - (32 \text{ ft/sec}^2)(32.2 \text{ sec}) = -530$ ft/sec.

3. p_b is a vector with $p_{bN}N$ the *north* component, and $p_{bE}E$ the *east* component, where

N and E are unit vectors; thus, $p_b = p_{bn}N + p_{bE}E$.

$$p_b = p_{bN}N + p_{bE}E.$$
$p_{bN} = m_N v_N = (1500 \text{ kg})(20 \text{ m/sec})$
   $= 3 \times 10^4$ kg m/sec;
$p_{bE} = m_E v_E = (2000 \text{ kg})(20 \text{ m/sec})$
   $= 4 \times 10^4$ kg m/sec.
$p_b = \sqrt{p_{bE}^2 + p_{bN}^2}$
   $= 5 \times 10^4$ kg m/sec.
$$v_a = \frac{p_b}{m_1 + m_2}$$
$$= \frac{5 \times 10^4 \text{ kg m/sec}}{3500 \text{ kg}} = 14.3 \text{ m/sec}.$$

4. $I = \dfrac{mL^2}{12} = \dfrac{(4 \text{ kg})(0.1 \text{ m})^2}{12}$

   $= 3.3 \times 10^{-3}$ kg m^2.

5. a The acceleration along the incline is $a = g\sin\theta$. The acceleration in the vertical direction is $a_{vert} = a\sin\theta = g\sin^2\theta$. The mass of the man is $\dfrac{w}{g} = \dfrac{160 \text{ lb}}{32 \text{ ft/sec}} = 5$ slug, and his apparent weight w' is, $w' = m(g - a_{vert}) = 120$ lb. Thus, $a_{vert} = \dfrac{-120 \text{ lb} + 160 \text{ lb}}{5 \text{ slug}} = 8$ ft/sec^2.
   b $a_{vert} = g\sin^2\theta$; hence, $\sin\theta = \sqrt{\dfrac{8}{32}} = \dfrac{1}{2}$, and $\theta = 30°$.

6. a $g = \dfrac{GM}{R^2}$

$$= \frac{(6.67 \times 10^{-11} \text{nt m}^2/\text{kg}^2)}{(2 \times 10^4 \text{m})(6.67 \times 10^{21} \text{kg})}$$

   $= 25$ m/sec^2.

b $U = \dfrac{GMm}{R}$

$$= -\frac{(6.67 \times 10^{-11} \text{nt m}/\text{kg}^2)(2 \text{kg})}{(2 \times 10^4 \text{m})(6.67 \times 10^{21})}$$

   $= -10^6$ joule.

7. a The volume of the block V_{bl} is 1 m^3; the buoyancy force F_b of the liquid is $F_b = \rho_{liq}gV_{displaced} = m_{bl}g$; thus, $m_{bl} = \rho_{liq}V_{displaced} = (2.04 \times 10^3 \text{kg/m}^3)(0.8 \text{ m}^3) =$

1.63×10^3 kg. Hence, $\rho_{bl} = \dfrac{m_b}{V_{bl}} =$

$\dfrac{1.63 \times 10^3 \text{kg}}{1\ \text{m}^3} = 1.63\ \text{kg/m}^3$.

b The additional buoyancy force is, $-x\,(1\ \text{m}^2)\,\rho_{liq}g = m_{bl}a_{bl}$; thus, $a = -12.25x$ m/sec^2.

8. $\dfrac{F_N}{A} = \dfrac{Y\Delta l}{l} = \dfrac{(2 \times 10^{12}\text{dy/cm}^2)(6\ \text{cm})}{3 \times 10^2 \text{cm}}$

   $= 4 \times 10^{10}\text{dy/cm}^3$.

9. a $n = (60\ \text{gm}\ O_2)\left(\dfrac{1\ \text{mole}}{32\ \text{gm}}\right) = 1.875$

mole; hence,
$$
\begin{aligned}
P_0 V_0 &= nRt \\
&= (1.87\ \text{mole}) \\
&\quad (8.31\ \text{joule/mole °K})(300°K) \\
&= 4.68 \times 10^3\ \text{joule and}\ P_0 V_1 \\
&= nR\,(400°K) \\
&= 6.24 \times 10^3\ \text{joule}.
\end{aligned}
$$

b $W = P_0\Delta V = 1.56 \times 10^3$ joule
$= 373.5$ cal and $\Delta U = nC_v\Delta T$
$= (1.87\ \text{mole})$
$\qquad (5.03\ \text{cal/mole °K})(100°K)$
$= 942$ cal.

c $Q = \Delta U + W + 942$ cal $+ 373.5$ cal $= 1315.5$ cal.

10. a The specific heat of water is
1 cal/gm °C; $\dfrac{\Delta Q}{\Delta t} = \dfrac{\Delta m}{\Delta t}\Delta T = (1\ \text{cal/gm}$
°C) $(20\ \text{gm/sec})(2°\ \text{C}) = 40$ cal/sec.

b $\dfrac{\Delta Q}{\Delta t} = -kA\dfrac{\Delta T}{\Delta x}$; thus,

$k = -\left(\dfrac{1}{A\dfrac{\Delta I}{\Delta x}}\right)\dfrac{\Delta Q}{\Delta t}$

$= -\left[\dfrac{1}{4\ \text{cm}^2\dfrac{(20°\ \text{C} - 100°\ \text{C})}{20\ \text{cm}}}\right]40\ \text{cal/sec} =$

$+2.5$ cal/cm sec^2° C.

11. $e = \dfrac{W}{Q} = \dfrac{10^4\ \text{joule}}{2 \times 10^4\ \text{joule}} = \dfrac{T_1 - T_2}{T_1} =$

$\dfrac{600 - T_2}{600} = \dfrac{1}{2}$; thus, $T_2 = 300°$K.

12. a $\mu = \dfrac{0.25\ \text{kg}}{10\ \text{m}} = 2.5 \times 10^{-2}\text{kg/m}$;
$\lambda = 20$ m; $f = 350$ Hz; $v = \lambda f = 7 \times 10^3$
m/sec; and $v = \sqrt{\dfrac{F}{\mu}}$. Hence, $T = \mu v^2 =$
$(2.5 \times 10^{-2}\ \text{kg/m})(7 \times 10^3\ \text{m/sec})^2 =$
122×10^4 nt.

b $f = \dfrac{v}{\lambda} = \dfrac{1}{\lambda}\sqrt{\dfrac{F}{\mu}}$. If F doubles,
f is multiplied by $\sqrt{2}$; thus, $f = 495$ Hz.

13. $E = \dfrac{F}{q} = \dfrac{\text{nt}}{10^{-4}\ \text{coul}}$
   $= 3 \times 10^4$ nt/coul.

14. $\mathcal{E} = -Bvl = -(2 \times 10^{-4}\ \text{weber/m}^2)\cdot$
$(100\ \text{m/sec})(10\ \text{m}) = -0.2$ volt.

15. The flux per turn is $\Phi = B\cdot S = \mu_0(N/l)IS$ and the total flux through the circuit is $N\Phi = \mu_0(N^2/l)IS = LI$, from Eq. 18-7. Thus, $L = \mu_0 N^2 S/l = \mu_0 N^2 Sl$.

16. $\dfrac{\text{Sin}\,\theta_1}{\text{Sin}\,\theta_2} = \dfrac{n_2}{n_1}$; $n_2 = 1.33$, and $n_1 = 1.63$.
Set $\theta_2 = 90°$, then $\sin\theta_1 = \dfrac{1.33}{1.63} = 0.816$;
thus, $\theta_1 = 54.7°$ is the critical angle.

17. $2dn = \left(N + \tfrac{1}{2}\right)\lambda$ and $\lambda = \dfrac{2dn}{N + \tfrac{1}{2}} =$

$\dfrac{2(3000\text{Å})(1.5)}{N + \tfrac{1}{2}} = \dfrac{9 \times 10^3\,\text{Å}}{N + \tfrac{1}{2}}$; thus,

$\lambda = 18{,}000$ Å $(N = 0)$ and $\lambda = 6000$ Å $(N = 1)$, etc. The only wavelength in the visible range is $\lambda = 6000$ Å, or a yellow-orange light.

18. $2d = \dfrac{v^2}{R} = N\lambda$ and $R = \dfrac{r^2}{N\lambda} =$

$\dfrac{(1.2\ \text{cm})^3}{100(6 \times 10^{-5}\ \text{cm})} = 240$ cm.

300° K	11
122×10^4 nt	12a
495 Hz	b
3×10^4 nt/coul	13
-0.2 volt	14
$\mu_0 N^2 Sl$	15
54.7°	16
yellow-orange	17
240 cm	18

TABLE A-1
The Carnot Cycle (See chapter 12.)

Process	State	Work	Heat Flow	Internal Energy Change ΔU	Entropy Change ΔS
1 (Isothermal)	P_1, V_1, T_1 to P_2, V_2, T_1 (a to b, Fig. A-1)	$W_1 = \int_{v_1}^{v_2} P\,dV$ $= nRT_1 \int_{v_1}^{v_2} \frac{dV}{V}$ $= nRT_1 \ln\left(\frac{v_2}{v_1}\right)$	$\Delta Q_1 = W_1$	0	$\frac{\Delta Q_1}{T_1} = \frac{W_1}{T_1}$
2 (Adiabatic)	P_2, V_2, T_1 to P_3, V_3, T_2 (b to c)	$W_2 = \int_{v_2}^{v_3} P\,dV$ $= \int_{v_2}^{v_3} \frac{K\,dV}{V^\gamma}$ $= \frac{K_1}{1-\gamma}(V_3^{1-\gamma} - V_2^{1-\gamma})$	0	$\Delta U_2 = -W_2$ $= -\Delta U_4$	0
3 (Isothermal)	P_3, V_3, T_2 to P_4, V_4, T_2 (c to d)	$W_3 = nRT_2 \ln\left(\frac{V_4}{V_3}\right)$	$\Delta Q_3 = W_3 < 0$	0	$\frac{\Delta Q_3}{T_2} = \frac{W_3}{T_2}$ $\Delta S_1 = \Delta S_3$
4 (Adiabatic)	P_4, V_4, T_2 to P_1, V_1, T_1 (d to e)	$W_4 = \frac{K}{1-\gamma}(V_1^{1-\gamma} - V_4^{1-\gamma})$ $= -W_2$	0	$\Delta U_4 = -W_4$ $= -\Delta U_2$	0

(See Fig. A-1, p. 95)

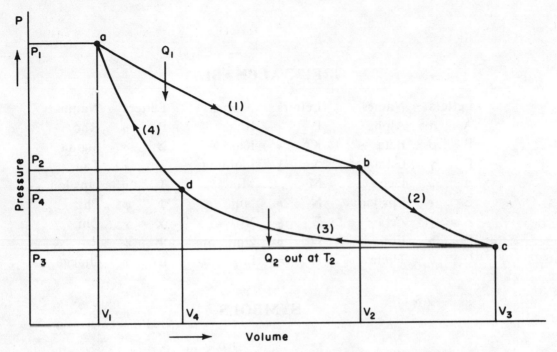

Fig. A-1.   The reversible process of the Carnot cycle described in chapter 12 (p. 46).

GREEK ALPHABET

Letters		Names	Letters		Names	Letters		Names
A	α	Alpha	I	ι	Iota	P	ρ	Rho
B	β	Beta	K	κ	Kappa	Σ	σ	Sigma
Γ	γ	Gamma	Λ	λ	Lambda	T	τ	Tau
Δ	δ	Delta	M	μ	Mu	Υ	υ	Upsilon
E	ϵ	Epsilon	N	ν	Nu	Φ	ϕ	Phi
Z	ζ	Zeta	Ξ	ξ	Xi	X	χ	Chi
H	η	Eta	O	o	Omicron	Ψ	ψ	Psi
Θ	θ	Theta	Π	π	Pi	Ω	ω	Omega

SYMBOLS

$=$, is equal to

$\neq$, is not equal to

$<$, is less than

$>$, is greater than

$\leq$, is less than or equal to

$\geq$, is greater than or equal to

log, logarithm

$\dfrac{d}{dx}$, derivative with respect to x

Σ, the sum of

$\equiv$, is identical to

P_1, "P sub 1"

$n \rightarrow \infty$, n approaches infinity

$\sqrt{n}$, square root of n

$\angle$ ABC, angle with vertex at B

Exp $x = e^x$, the exponential function of x

Δx, increment of x

f(x), function of x

$\int$, integral of.

$\bar{a}$, the average value of a

$\propto$, is proportional to

TABLE A-2
Physical Quantities

Quantity	Symbol	Unit
Acceleration	a	m/sec^2
Angular acceleration	α (alpha)	rad/sec^2
Angular displacement	θ (theta)	rad or degree (°)
Angular frequency	ω (omega)	rad/sec
Angular momentum	L	kg m^2/sec, slug ft^2/sec
Angular velocity	ω (omega)	rad/sec
Area	A or S	m^2
Capacitance	C	farad
Charge	q	coul
Current	i, I	amp
Density	ρ (rho)	kg/m^3
Dipole moment		coul m
Electric field	E	volt/m
Electric flux	Φ (phi)	volt m
Electric potential	V	volt
Electromotive force (emf)		volt
Entropy	S	joule/°K
Force	F	nt, dy, lb
Frequency	f	Hz, cycles/sec
Gravitational force	g	nt/kg
Heat	Q	joule
Inductance	L	henry
Kinetic energy	K	joule
Length	l, L, r	m, cm, ft, in.
Magnetic field	B	amp m
Magnetic flux	Φ (phi)	weber, volt sec
Magnetic induction	B	weber/m^2
Mass	m	kg, gm
Moment of inertia	I	slug ft^2
Momentum	p	kg m/sec^2
Period	T	sec
Permeability	μ (mu)	henry/m
Permittivity	ϵ (epsilon)	farad/m
Potential energy	U	joule
Power	P	watt, joule/sec, hp, ft lb/sec
Pressure	P	nt/m^2
Resistance	R, Ω (omega)	ohm
Resistivity	ρ (rho)	ohm m
Temperature	T	degree (°)
Time	t	sec
Torque	τ (tau)	nt m
Velocity	v	m/sec
Voltage	V	volt
Volume	V	m^3, cm^3
Wavelength	λ (lambda)	m, μ (micron), Å (Angstrom)
Work	W	joule

TEMPERATURE CONVERSION SCALES

The three most common scales for measuring temperature are: **Fahrenheit** (°F), **Celsius**, or *centigrade*, (°C), and **Absolute**, or *Kelvin*, (°K). The temperature scale conversion formulas are,

$$°F = (9/5 \times °C) + 32$$
$$°C = (°F - 32) \times 5/9$$
$$°K = 273 + °C$$

Boiling point: 212° F, 100° C, 373° K
Freezing point: 32° F, 0° C, 273° K
Lowest possible temperature: −459° F, −273° C, 0° K

SYSTEMS OF UNITS

System	Length	Mass	Time	Force
mks (meter-kilogram-second)	meter	kilogram	second	newton
cgs (centimeter-gram-second)	centimeter	gram	second	dyne
fps (foot-pound-second)	foot	slug	second	pound

COMMON UNITS OF MEASUREMENT

Units of length:

1 meter (m) = 39.37 in.
1 centimeter (cm) = 10^{-2} meter = 3.9×10^{-1} in.
1 kilometer (km) = 10^3 meters = 6.2×10^{-1} mile
1 millimeter (mm) = 10^{-3} meter = 4×10^{-2} in.
1 mile (mi) = 5280 ft = 1760 yd = 320 rd
1 rod (rd) = 5.50 yd = 16.5 ft
1 yard (yd) = 3 ft
1 foot (ft) = 12 inches (in.)
1 Angstrom (Å) = 10^{-8} cm
1 micron (μ) = 10^{-4} cm

Units of mass:

1 slug = 1 pound/g, where g = 980 cm/sec^2 = 9.8 m/sec^2 or 32 ft/sec^2
1 kilogram = 10^3 grams

Units of force:

1 dyne (dy) = 1 gm cm/sec^2 (cgs)
1 newton (nt) = 1 kg m/sec^2 (mks)
1 pound (lb) = 1 slug ft/sec^2 (fps)

Units of work and energy:

1 electron volt (eV) = electron change × 1 volt = 1.6×10^{-19} joule
1 joule = 1 nt m = 10^7 erg
1 erg = 1 dy cm or ft lb.

Units of Power:

1 watt = 1 joule/sec = 10^7 erg/sec
1 ft lb/sec = 1/550 horsepower (hp)

Units of heat:

1 kilcalorie (kcal) = 1000 calories (cal)
1 British thermal unit (BTU) = 252 cal = 0.252 kcal
1 cal = 4.19 joule

Electrical units:

1 volt = 1 joule/coul
1 farad = 1 coul/volt
10^6 farad = 1 microfarad (μf or μfd)
10^{-12} farad = 1 micromicrofarad ($\mu\mu$f or $\mu\mu$fd)
1 watt (w) = 1 joule/sec = 1 (joule/coul)(coul/sec) = 1 volt amp
1 megawatt (Mw) = 10^6 watts
1 kilowatt (kw) = 10^3 watts
1 milliwatt (mw) = 10^{-3} watt
1 watt sec = 1 joule
1 kilowatt hour (kwh) = 3.6×10^6 joule
1 ohm (Ω) = 1 volt/amp
1 megohm (M or meg) = 10^6 ohms
1 kilohm (K) = 10^3 ohms

Units of magnetism and inductance:

1 weber/m^2 = 1 nt sec/coul m = 1 nt/amp m = 10^4 gauss
1 weber = 1 volt sec = 10^8 maxwells
1 henry = 1 volt sec/amp = 1 weber/amp

Units of circular measure:

1 degree = 1.745×10^{-2} radian
1 radian = 57.30 degrees

PHYSICAL CONSTANTS

Permittivity constant e_0 = 8.854×10^{-12} coul^2/nt m^2
$\qquad$ = 8.85×10^{-12} farad/m
Proportionality constant $1/4\pi\epsilon_0$ = 9.0×10^{-9} nt m^2/coul^2
Permeability constant μ_0 = $4\pi \times 10^{-7}$
Speed of light c = 3.00×10^8 m/sec
Stefan-Boltzmann constant σ = 5.67×10^{-8} watt/m^2 °K^4
Gravitational (Newton's) constant G = 6.67×10^{-11} nt m^2/kg^2 = 6.67×10^{-8}
$\quad$ dy cm^2/gm^2
Universal gas constant R = 8.315×10^7 (dy/cm^2) cm^3/mole °K
$\quad$ = 8.315×10^7 erg/mole °K = 8.315 joule/mole °K
$\quad$ = 1.99 cal/mole °K = 0.082 liter atmosphere/mole °K.
Equatorial radius of the earth = 6.378×10^6 m = 3963 mi
Polar radius of the earth = 6.357×10^6 m = 3950 mi
Volume of the earth = 1.087×10^{21} m^3 = 3.838×10^{22} ft^3
Mass of the earth = 5.983×10^{24} kg

COMMON NUMERICAL VALUES

π = 3.1416

$\sqrt{\pi}$ = 1.77245

Exponential function (exp) e = 2.71828

$\sqrt{e}$ = 1.64872

$\sqrt{2}$ = 1.414

$\frac{1}{2}\sqrt{2}$ = 0.707

$\sqrt{3}$ = 1.732

$\frac{1}{2}\sqrt{3}$ = 0.866

$\frac{1}{3}\sqrt{3}$ = 0.577

Absolute temperature scale (Kelvin): See under **Temperature scales.**

Acceleration: A vector quantity, the time-rate change of velocity, 8

  Angular: The time-rate change of angular position, 22

  Average: The total displacement divided by the total time elapsed, 8

  Centripetal: The acceleration experienced by a body moving in a circle, 8

  Of gravity: Uniformly accelerated motion of a free-falling body. The universal gravitational constant $g = 9.8$ m/sec^2 or 32 ft/sec^2, 6, 10

  Tangential: The acceleration experienced by a particle moving in a circle of constant radius and whose angular acceleration is changing with time, 22

Actual mechanical advantage: See under **Mechanical advantage.**

Adiabatic process: A process in which no heat is either absorbed or given off by a system to its environment, 42, 44, 46, 48

Aerodynamic lift: Lift caused by unequal pressures on two regions of a body moving in a fluid, 36

Alternating current: See under **Current.**

Alternating-current generator: A generator that provides current which varies sinusoidally with time, 74, 76

Ampere (amp): Mks unit of current, 62

Ampere's law, 68

Angstrom (Å): Unit of length, 81, 83.

Angular acceleration: See under **Acceleration.**

Angular momentum: See under **Momentum.**

Angular velocity: See under **Velocity.**

Antinode: A point on a standing wave pattern for which the amplitude of oscillation is a maximum.

Archimedes' principle: A body immersed in a fluid is lifted by a force equal to the weight of the fluid displaced by the body, 34.

Atwood machine: Two weights connected by a string which passes over a pulley, 9

Average speed: See **Speed, average.**

Average velocity: See under **Velocity.**

Beats: When two sound waves of slightly different frequencies are heard together, the amplitude of the resulting sound waves varies sinusoidally with a frequency equal to the difference in the two frequencies. This is called the *beat frequency,* 52

Bernoulli's equation: An equation of fluid dynamics equivalent to the conservation of energy equation, 36

Biot-Savart law: An equation used to calculate the magnetic field due to a current distribution, 66 (Eq. 17–10a).

British thermal unit (BTU): A unit of heat energy. The amount of heat necessary to raise the temperature of 1 standard pound of water from 63°F to 64°F, 44

Bulk modulus: The ratio of the pressure of a body to the fractional change in volume, 52

Buoyancy force: See under **Force.**

Calorie: Unit of heat, 44

Calorimetry: Methods of measuring the absorption or radiation of heat, or of determining specific heats, 42, 44

Capacitance: The ratio of the charge on a conductor to the electric potential. The unit of capacitance is the *farad*, 60, 64, 74, 76

Capacitive reactance: See **Reactance.**

Capacitor: A device that provides capacitance in a circuit, 63, 74, 76

Capacity,

  Electric: See **Capacitance.**

  Heat: See under **Heat.**

Carnot cycle: A thermodynamic system is carried through a set of four reversible, isothermal, and adiabatic processes, the end of which is the original state of the system. This cycle is used to study the efficiencies of heat engines and the second law of thermodynamics, 46, 48

Celsius scale (centigrade): See under **Temperature scales.**

Center of gravity: See under **Gravity.**

Center of mass: The theoretical point at which the mass of an object is concentrated. If an external force is applied at the center of mass of a body, this body would move as though the force were applied to a particle at the point where the center of mass is located, 18, 20

Centimeter-gram-second (cgs): System of units, 2

Centripetal acceleration: See under **Acceleration.**

Charge: An elementary property of matter. There are two kinds of charge, positive and negative. Charges of like sign repel one another, and charges of unlike sign attract one another, 54

Circuit: A circuit is a system of elements (resistors, capacitors, batteries, inductors, etc.) connected to one another, 62, 64, 74, 76

Coefficient of linear expansion, 42

Coefficient of restitution, 20

Coefficient of sliding friction, 12

Coefficient of volume expansion, 42

Collision, perfectly elastic: See **Perfectly elastic collision.**

Compression stress: See under **Stress.**

Condenser: See **Capacitor.**

Conduction, heat: See under **Heat.**

Conductivity, thermal: The measure of a substance's ability to pass heat energy, 44

Conductor, electric: A substance that permits electrical charges to be transferred from one point to another, 62

Conservation of angular momentum: See under **Momentum.**

Conservation of energy: See under **Energy.**

Conservation of linear momentum: See under **Momentum.**

Conservative force: See under **Force.**

Continuity, equation of: The mass of fluid flowing into a volume equals the mass of fluid flowing out of the volume, 36 (Eq. 9-5).

Convection, heat: See under **Heat.**

Coulomb (coul): Mks unit of charge. Equal to the charge carried past a point by a 1 ampere of current, 54, 56

Coulomb's law: The attractive force between two charges is

Frictional: The force exerted by a body sliding on a surface. A frictional force always opposes motion, 12

Gravitational, 30

Magnetic: The force on a charge moving in a magnetic field, 66

Nonconservative: A force which is not conservative, 14

Free body diagram: A diagram that shows all forces acting on a body within a reference frame. Free body diagrams are used in problems involving Newton's Laws, 10, 12

Frequency: The number of revolutions of a wave or vibration per unit time, 32, 50, 74, 76

Angular, 50, 74, 76

Fundamental: The lowest standing wave frequency possible in an oscillating system, 50

Frictional force: See under **Force.**

Galilean transformation (nonrelativistic), 86

Galvanometer: Instrument for detecting current flow, 64

Gamma rays: High-frequency radiation.

Gas, ideal, 40

Gauss: Unit of magnetic induction, 68

Gauss's law, 54

Generator: See **Alternating-current generator.**

Gravitational force, 30

Gravitational potential energy, 30

Gravity (gravitational force),

Acceleration of: See under **Acceleration.**

Center of: A point within a body through which the gravitational force acts upon the whole body.

Specific: The ratio of the density of a material to the density of water, 34

Harmonic motion,

Rotary (RHM): Motion along the arc of a curve in which the angular acceleration is proportional to the angular displacement.

Simple (SHM): A vibrational, linear motion about an equilibrium point in which the acceleration is always proportional to the displacement from the equilibrium point and directed to it, 32

Heat: Energy in transit from a higher to a lower temperature, 42, 44, 46, 48

Capacity: The ability of a substance to store heat energy, 42

Conduction: The flow of heat from a hotter to a cooler section of a body, 44

Convection: The movement of heat resulting from an agent, such as air or water, 44

Radiation: The transfer of heat energy by electromagnetic waves, 42

Specific: Heat capacity per unit mass, 42

Transfer: The movement of heat from one place to another by conduction, convection, or radiation, 42, 44

Heat capacity: See under **Heat.**

Heat conduction: See under **Heat.**

Heat convection: See under **Heat**

Heat engine: See **Carnot cycle.**

Heat radiation: See under **Heat.**

Heat transfer: See under **Heat.**

Henry: The unit of inductance, 72

Huygen's principle: Every point on a three-dimensional wavefront acts as a point source for secondary waves, 78

Hydrostatics: See Fluid statics.

Ideal gas, 40

Equation of state for, 40

Ideal mechanical advantage: See under **Mechanical advantage.**

Image,

Real: An image formed by light rays meeting at a point, 78

Virtual: An image formed by extended rays meeting at a point, 78

Impedance: The total opposition to the flow of charges in a circuit in which the current alternates or in any way varies, including resistance and reactance. Impedance is measured in ohms, 76

Index of refraction: See under **Refraction.**

Induced current: Current developed in a coil of wire resulting from varying the magnetic field about the coil, 70

Inductance: See Induced current.

Mutual: The production of an emf and current in a circuit lying in the field of another circuit in which the emf is varying, 72

Self-: The effect of mutual inductance between the turns of wire in a coil, 72

Inductive reactance: See **Reactance.**

Inductor: A coil of wire that provides inductance to a circuit, 64, 74, 76

Inertial reference frame: See under **Reference frame.**

Instantaneous velocity: See under **Velocity.**

Interference: The result of coherent waves adding by superposition, 82

Internal energy, 46

Irreversible process: See **Reversible process.**

Isotherm: A line on a pressure vs temperature curve, which represents an isothermal process.

Isothermal process: A process in which the same temperature is maintained throughout, 46, 48

Joule: Mks unit of energy, 16

Joule heating: The heat per second produced by a current through a resistor, 62, 64

Kelvin temperature scale: See **Absolute** under **Temperature scales.**

Kepler's laws: Three laws that describe the motion of planets orbiting about the sun, 30

Kinematics: The study of the motion of bodies, 6, 8

Kinetic energy: See under **Energy.**

Kirchhoff's laws, 64

Lens: A transparent material shape to *refract* light in certain ways. The faces of a lens are usually sections of spheres, 80, 84

Focal length: The distance from the focal point to the center of a lens, 80

Focal point: The point at which parallel rays of light incident on a lens are focused, 80

Lens equation, 78

Lensmaker's equation, 80

Lenz's law: The direction of the induced current in a circuit is always such that the electric flux it produces tends to oppose the change in the magnetic flux through the circuit, 70.

Light: An electromagnetic wave traveling at 3×10^8 m/sec, 78, 80, 82, 84

Light wave, 78, 80, 82, 84

Linear momentum: See under **Momentum.**

Wavelength: The distance between points of equal phase on a wave, 50

Weber: Unit of magnetic flux, 72

Weight: The effect of the force of gravity upon a body, 10

Weight density: See under **Density.**

Wheatstone bridge: A device used to compare resistances, 64

Work: The product of force and distance, 14, 16, 46

Work energy theorem: See under **Energy.**

X ray: Electromagnetic radiation produced by bombarding certain material, especially metals, with cathode rays.

Young's modulus: The ratio of compressional stress to the fractional change in length of a material, 38